Practical Health and Safety Management for Small Businesses

To Leslie

Practical Health and Safety Management for Small Businesses

Jacqueline Jeynes

OXFORD AUCKLAND BOSTON JOHANNESBURG MELBOURNE NEW DELHI

Butterworth-Heinemann
Linacre House, Jordan Hill, Oxford OX2 8DP
225 Wildwood Avenue, Woburn, MA 01801-2041
A division of Reed Educational and Professional Publishing Ltd

A member of the Reed Elsevier plc group

First published 2000

British Library Cataloguing in Publication Data
Jeynes, Jacqueline
Practical health and safety management for small businesses
1 Industrial safety – Great Britain 2 Small business – Great Britain – Management
I Title
363.1'1'0941

Library of Congress Cataloguing in Publication Data
Jeynes, Jacqueline.
Practical health and safety management for small businesses/Jacqueline Jeynes.
p. cm.
Includes index.
ISBN 0 7506 4680 2
1 Industrial hygiene 2 Industrial safety
3 Fire prevention 4 Small business I Title.
HD7261 .J49 2000
658.4'08–dc21 00–033745

ISBN 0 7506 4680 2

Composition by Genesis Typesetting, Laser Quay, Rochester, Kent
Printed and bound in Great Britain by Biddles Ld
www.biddles.co.uk

Contents

Preface

During the last six years, I have looked closely at the impact of changes to health, safety and fire legislation on very small firms, and have brought many of the problems and concerns to the notice of the legislators. Research carried out for my PhD also highlighted the need for a more practical approach to guidance that starts with the business itself, rather than what the law requires.

This guide tries, therefore, to take a straightforward, practical approach to identifying, organizing, and managing the health, safety and fire risks in the business. The crucial point is that while it does not guarantee that you have covered every aspect of the legislation that applies to your business, it does establish a system that you can maintain, add to, and keep up to date as your needs and the regulations change. If nothing else, it has made you look very closely at how the business is organized, and you now have a file of evidence to show anyone who wants to see it!

Jacqueline Jeynes

Acknowledgements

I would like to thank the following organizations for allowing me to use illustrations of their business, or for providing information and assistance in the preparation of this book:

> Carmichael International; Dane Architectural; Dove Stores; Fleet Services Ltd; Fosse ME Ltd; L.J. Narraway – Butchers; Natural Break; Opal Services; F. Parr Ltd; St John's Dental Repairs; Station Café; The Studio; Tiltridge Vineyard; Twinkle-toes – chiropodist; Vulcascot; Welton Packaging Ltd; plus the unknown Tree Feller who gave permission to use the photogaphs!

I would, of course, also like to thank all the individuals and organizations involved with health, safety and fire risk assessment that I have worked with over the last few years, including Professor Richard Booth, David Smith, Stephen Fulwell, and David Morris from HSE.

Finally, I would like to dedicate this book to my husband, Leslie, who made sure I completed it on time!

Glossary of main terms

This guide has been written with the intention of avoiding the use of technical terms and 'jargon' as much as possible. The following are the main terms used.

- Hazard – something with the potential to cause harm or injury
- Risk – the likelihood that it will actually cause harm or injury
- Risk assessment – the process of identifying hazards and assessing the severity of harm and likelihood it will occur
- HSC – Health and Safety Commission are the national body with the responsibility for considering health and safety issues, and where the law may need to be amended to provide better or further protection for workers and others in the workplace
- HSE – the Health and Safety Executive answers to the HSC, acting as inspectors to ensure that the law is adhered to and producing relevant guidance
- COSHH – the control of substances that might be hazardous (that is, with the potential to cause harm or injury) when used or stored, or disposed of. The substances can be liquids, gases, fumes, dusts, and can be absorbed through direct contact with the skin, through breathing in, through swallowing, and via other means such as through puncture wounds
- VDUs – the screens people use at computer terminals
- Manual handling – the action of handling large/heavy/awkwardly shaped/ compact/uneven/ or sharp edged objects (including people or animals). It relates to lifting, pulling, pushing or carrying these objects, and the potential damage it can do to people if they handle things incorrectly. This can be lower back injuries, injuries to upper parts of the body and limbs, and other injuries associated with dropping the object

Part A

Context

Introduction

Where to start?

Dealing with health and safety in the current business climate is extremely difficult for many firms, but particularly small firms where resources are at a premium. Everyone knows they have to do something about health and safety but many are not quite sure what to do or where to start, so put off any action until prompted by an event/incident/person that acts as a catalyst.

We will assume for now that something has prompted you to pick up this pack, and to take the first step towards organizing and dealing with aspects of health and safety in your workplace, and will briefly consider how this might colour the way you tackle the activities.

Motivation – why are you doing this at this time?

An inspector has visited your premises and identified things you need to address

This might be seen as positive or negative, depending on the inspector's findings and whether any enforcement action has been threatened or taken. However, the main motivation is probably to comply with the inspector's demands. But, as a business owner you know that it is more effective in the long term to take an overall view of a problem than just try to deal with parts of it piecemeal, therefore it pays to look at the big picture, as suggested in this guide, to

- identify current problems;
- deal with them appropriately;
- keep records to show to inspectors or others the actions taken;
- and keep control of the situation in the future.

If a Local Authority or Health & Safety Executive (HSE) inspector has visited your premises, and taken some sort of formal action, he or she will be required to follow it up. You can expect further visits well into the future, so it is vital that you can show you are in control of the situation.

An accident or incident or near-miss has occurred

While you are required to report serious accidents and incidents that result in injury, this will not always result in a formal investigation being carried out by HSE Inspectors. However, it is vital that you carry out your own internal investigation to ensure the same thing does not happen again, particularly injury to workers or customers. Accident and Emergency departments in hospitals around the country display posters from local solicitors who specialize in litigation for victims of accidents.

Main requirements of RIDDOR (Reporting of Injuries, Diseases and Dangerous Occurrences Regulations)

You should report the incident if:

- an employee or self-employed person working on your premises is killed, or suffers a major injury;
- a member of the public is killed or taken to hospital from your premises;
- an employee or self-employed person working on your premises suffers an injury that results in more than 3 days absence from their normal work;
- a doctor notifies you that your employee suffers from a reportable work-related disease;
- something happens that could clearly have resulted in major injury, but was avoided in this instance.

Check with HSE or Local Authority if not sure whether an incident should be reported.

Just as important is the 'near-miss' incident, where if it wasn't for the action of someone the result would have been disastrous, or the damage-only incident. Clearly as a business, it is in everyone's interest to ensure conditions in the workplace do not easily lead to accidents and incidents, and that adequate controls are in place.

The insurance broker or provider has asked to see evidence of the health and safety policy

This might lead on from the section above or might just as easily be a normal part of their procedure. As a very small firm, you are unlikely to get an on-site visit from your insurers, so getting the information right on the proposal form is crucial for you and them in the case of a claim. While protection of property and people is their prime concern, more and more insurance companies are looking for evidence that clients are actually managing health, safety, fire and security properly. There is also some evidence that premiums may be reduced if you can show them that you are in control, so this is certainly a positive motivator!

A major client or tender for contract requires it

The majority of small firms provide goods and services to large organizations that generally have more formally structured systems in place to deal with health and safety issues. There is increasing pressure on small suppliers to demonstrate that their own management system for health and safety is adequate in order to gain contracts. It may just require you to complete a checklist or questionnaire provided by the client, or submission of your own documents.

However, as we start the 21st century, there is even greater pressure internationally for firms to have a formal Occupational Health and Safety Management System (OH&SMS) in place that conforms to a specified standard and is given a certificate to say so by an outside body (third party certification). This could be similar to the ISO 9000 and 14000 series for Quality and Environment. While this guide is not structured in the same way as these systems, or the British Standard BS 8800 Guidelines, the evidence you generate while working through it will provide the basis for such a system if you need to organize it in that way in the future.

Employees or workers have asked for some action

You may have workers at various levels throughout the organization that have asked for action on specific health and safety issues as problems emerge. As health and safety is clearly about people and how they behave or react in different situations, it is important to get it right. Apart from ethical considerations, the corporate and personal liability of the business owner if anything goes wrong is significant, and with the requirement to consult with workers on health and safety issues, there is clear evidence that involving them leads to more positive attitudes company wide.

> Whatever size of firm you have, you have to have some procedure in place for consulting with employees on issues of health and safety. In very small firms, this will probably just mean talking to all workers together at the same time. In a larger business, you might need to consult through a smaller group of people, including elected representatives of the workforce. In unionized firms, there will probably already be such a system in place. Consulting means discussing health and safety issues directly with the people who are affected by them – not just passing details on of decisions already made without their input.

There are business benefits from taking action

This probably encompasses all the items listed above, as getting it wrong in these areas can constitute a huge cost to the business. However, it might also be that specific concerns have prompted further action, such as

- a review of sickness absence figures;
- increases in production of scrap as equipment or machinery becomes old or worn;
- the need to purchase new types of equipment;
- changes in the content or structure of materials – for example manufacturers of chemical substances change the properties of the product because of COSHH requirements (Control of Substances Hazardous to Health);
- changes in customer demand or expectations.

As noted previously, if you establish an approach to dealing with risks to the business in a holistic way, then it can be applied to a range of issues and concerns as they arise. There is no specified format or structure for much of the material you produce while working through this Guide, as it will be personal to your business. However, the way you collect and order the material should provide people outside and inside the business with evidence that you

(1) have identified the current situation;
(2) have identified problem areas that need to be addressed;
(3) have dealt with them appropriately;
(4) can demonstrate what actions have been taken;
(5) are in control of the situation.

There are some checklists included in this guide, but it is structured mainly along the lines of asking questions and identifying areas for which you might need to take further action. Much of the information you will already have, either as internal records, official documents, procedures and Training Manuals, leaflets or brochures. It is suggested that you collect all the evidence produced as you work through the Guide, and store it in a box file (or similar) with some means to divide sections where necessary – see later sections.

While marketing management traditionally has its marketing mix based on the interaction between the four Ps – Product – Price – Place – Promotion – it seems only fair that Health and Safety Management should have its own prompt list of elements to think about. This Guide is, therefore, based around a list of eight Ps that represent the main elements related to managing health, safety, fire and other risks in the workplace, and the way they interact with each other.

The eight Ps of health and safety risk management

*Premises	*Product	*Process	*Procedures
*People	*Purchasing	*Policy	*Protection/Prevention

These are not listed in any particular preference order, although it is easier to start from the most visible and tangible elements of what the business makes and sells, than the policy and strategy that leads or supports it.

We shall start to look more closely at your own business now, and the way it operates, starting with the elements that you are already familiar with.

Chapter 1

Your business

1.1 The context

It is valuable to set out the context in which you are working so that others can see how and why the documents have been produced. This is, of course, the easy bit as it is **your** business, and if you don't know about it then who does!

You can use the checklist on page 9 as a proforma which you fill in, or as a prompt list for collecting together and organizing data from various sources.

The checklist is intended to help you reflect on where you are at present, the way you organize and manage the business generally. It is the sort of information you would put together as a summary or introduction to a business plan when applying to banks or other financial institutions for funding, or indeed as part of an annual report for shareholders. Depending on the size, type and structure of your particular business, it will be as brief or as detailed as necessary to give an outsider a picture of the context in which the risks are being managed, without getting bogged down in too much detail which will be included in later sections.

1.2 The product or service

To complete this section about your business, it is important to include details of the product or service itself. The following list includes the main elements that you need to include in the description of a product, but is not necessarily exhaustive.

- particular features of the product that make it recognizable to the customer as yours;
- physical and intellectual properties;
- the main materials used in its production;
- its environmental impact, or descriptions of how 'environmentally friendly' it is;
- packaging used, including the materials used;

Checklist 1.1 Details about the business

1. Name of business.
2. Legal structure (include copy of any legal agreement).
3. Are there shareholders? If so, how many?
4. If so, what are their expectations regarding managing health, safety, fire and other risks?
5. What type of industry sector is your business in? Try to be specific in your description, rather than just 'service' or 'engineering'.
6. What do you actually make, or provide as a service? List the full range of products and services, and include any promotion or information literature you have.
7. How long has the business been established?
8. How is the business organized – include a simple flow chart to show movement of work through the system, and consider the following questions:
   - Is work activity based on one site?
   - Is work carried out on other people's premises?
   - How many buildings, yards, etc. make up the business?
   - Do staff have to travel between buildings/sites/clients for work purposes? If so, how is this organized?
   - Do products have to be transferred between sites or buildings?
9. Do customers come onto your site/premises?
   If so, is it free or controlled access during business hours? Include any points you think are relevant to health and safety risks in your business. Include copy of the Visitor's Book or other records.
10. What have been the most significant changes in the way the business is organized since it started?
11. How many people work for you in total, either as employees or on another basis?

full time permanent	part time permanent
full time temporary	part time temporary
on specified short-term contract	work experience
freelance self-employed	voluntary
other (please specify)	

12. What are the main lines of authority and responsibility between people in the business? Include an organization chart.

- range of uses for the product;
- how it is disposed of when obsolete;
- what sort of pre- or after-sales service you offer customers;
- any maintenance required or offered.

You could see this as the promotion bit, as it is an opportunity to clarify the exact nature of your product or service, both for yourself and for the external bodies that will be looking closely at the results of this activity. In particular, it helps you to give a more precise picture of the context in which the business operates, especially for enforcers or insurers who may not be totally familiar with the type of product or service you provide. The following list is more appropriate for a service:

- pre-sales service provided for customers or clients;
- where and how customers can access the service;
- information and guidance available to customers;
- intellectual properties;
- skills and qualifications of staff providing the service;
- control systems in place;
- after-sales support available.

Of course, it also helps to explain and justify how and why some of the processes operate as they do. After working through this Guide, it may also serve as a basis for revising some of these features if your evaluations suggest that changes need to be made to either the product itself or the way it is made or sold. Include technical literature and specifications where they exist, plus examples of promotional literature you use.

Chapter 2

The premises

The easiest starting point is to look closely at where the business operates; the internal and external structure and shape of the buildings, storage areas, car parks, etc. and the way it is utilized. As we know, premises are not always ideal for what we want to use them for, and quickly become overcrowded or outgrown as the business becomes established and expands. In addition, there may be restrictions on what can or cannot be altered, and the basis of ownership presents its own problems.

2.1 Site plan

Plan A – Outline plan of the site. Safety and security

Start with a diagram of the site and premises. Whether you own or lease the premises, you should have a copy of the boundary lines for both land and buildings, including shared or communal areas. If not, then try to get a copy from either your landlord, solicitor, or the Local Authority Planning Office.

If you have a site plan, have two or three photocopies prepared, preferably enlarged to make it easier to work on. If you do not have one, then draw an outline plan for yourself – more or less to scale will be sufficient! – showing the following features:

- external boundary of land;
- roads and footpaths;
- access routes for vehicles and pedestrians;
- external walls of all buildings, including sheds or outbuildings;
- car park areas;
- delivery areas.

Alternatively, use one of the Sample plans 1–4 included on the next four pages, and amend them where necessary to fit your own situation.

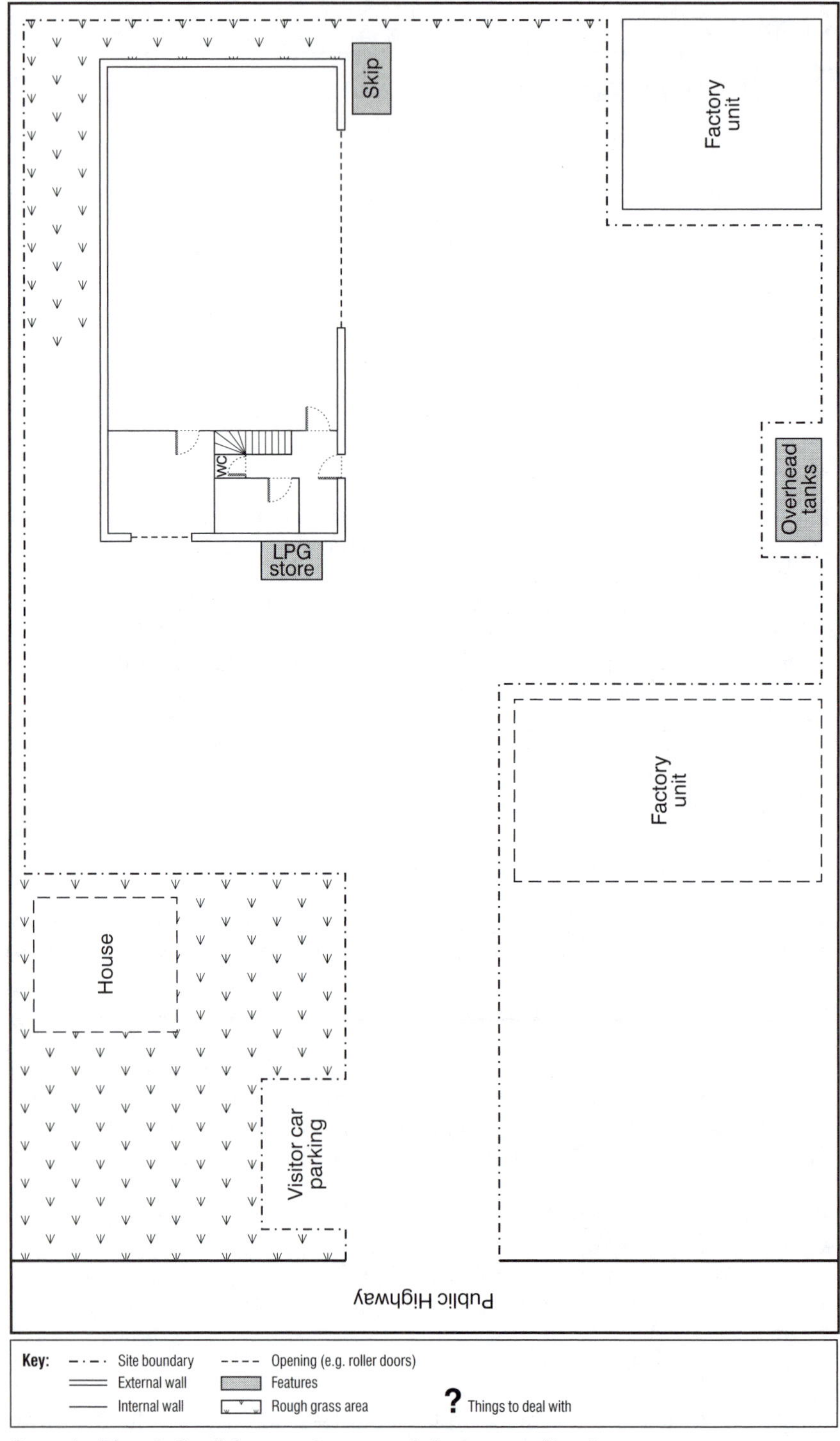

Sample Plan 1 Small factory site or workshop space: Plan A

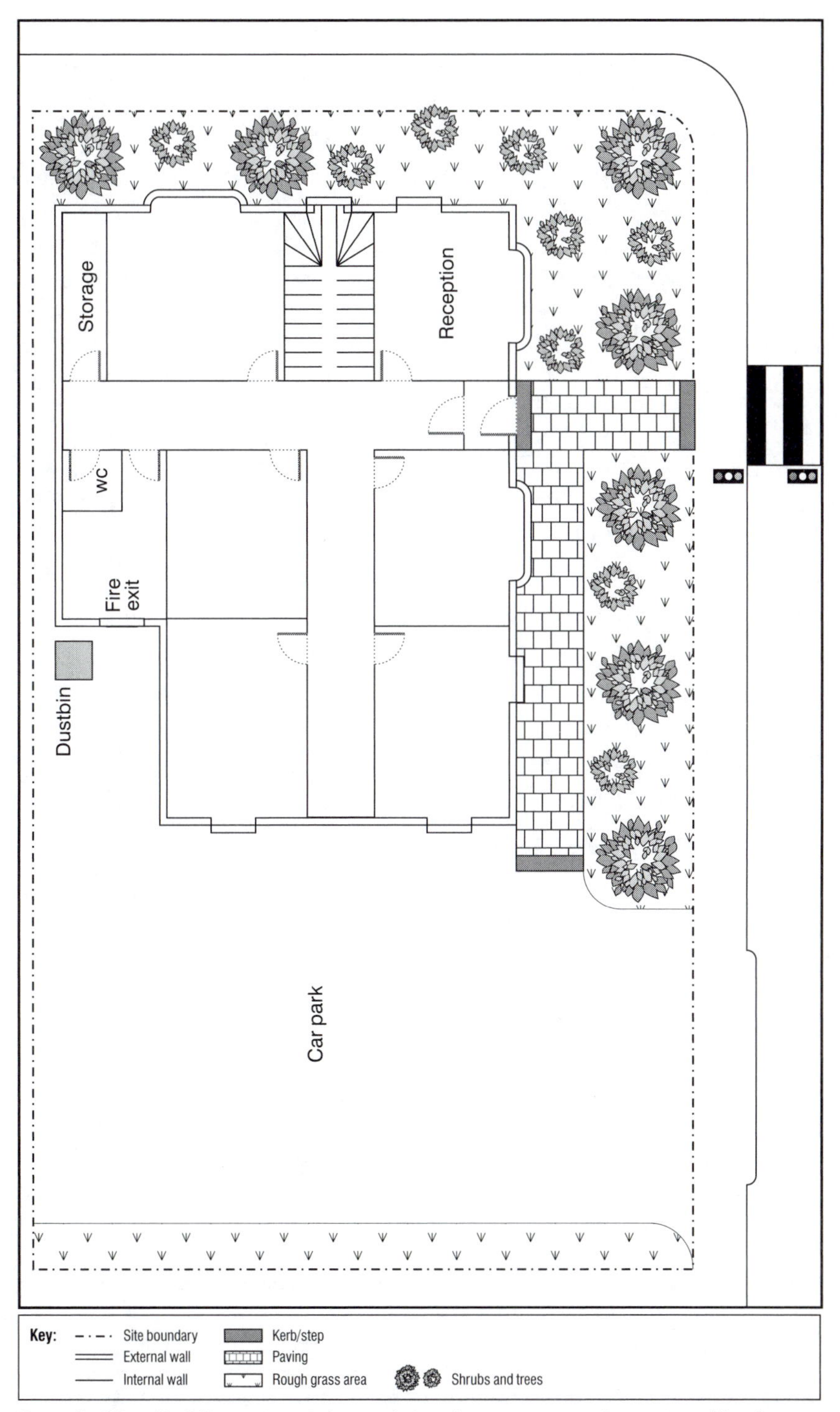

Sample Plan 2 Office suite, such as solicitors' or accountants' practices: Plan A

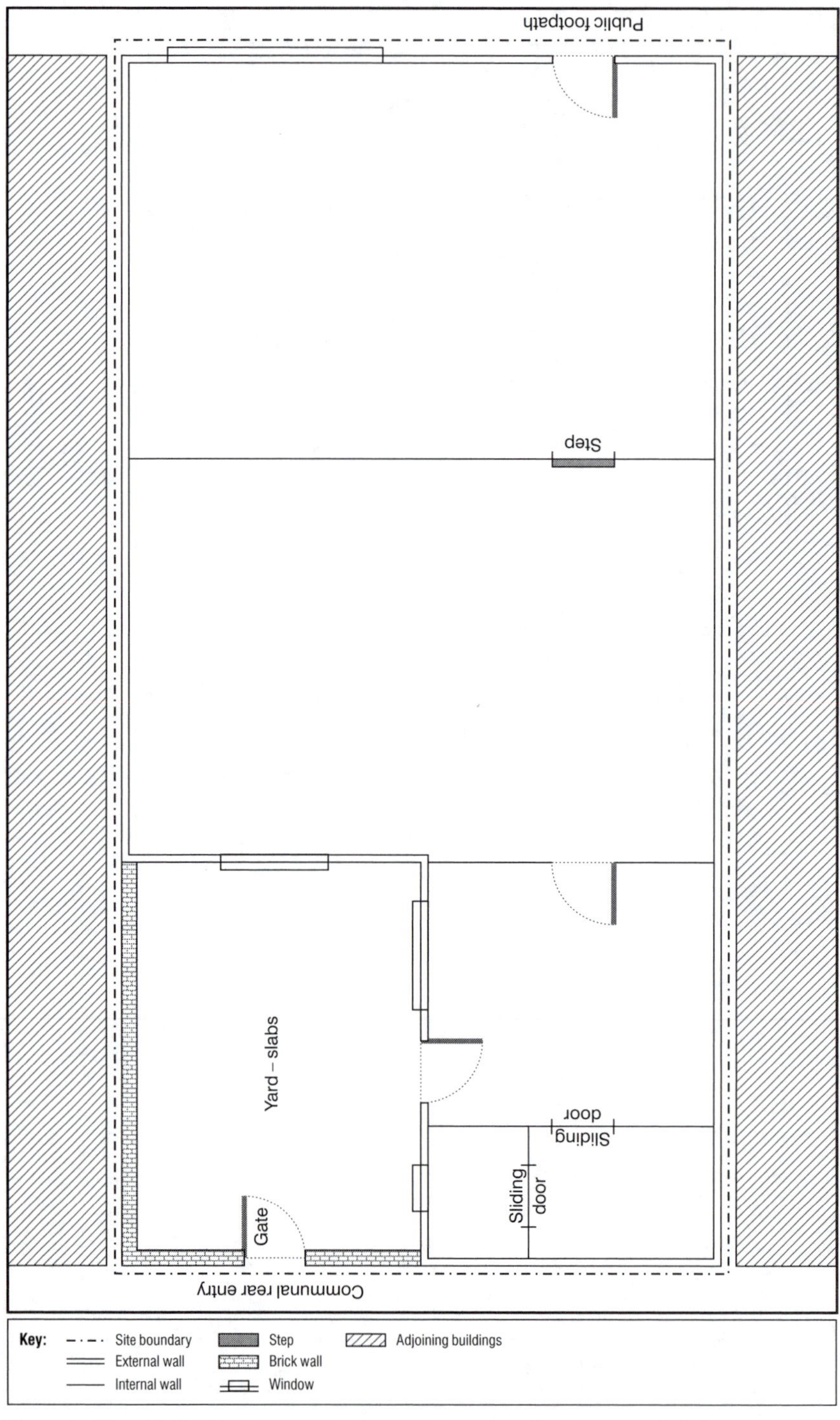

Sample Plan 3 Small town centre shop, or premises for opticians, hairdressers, etc.: Plan A

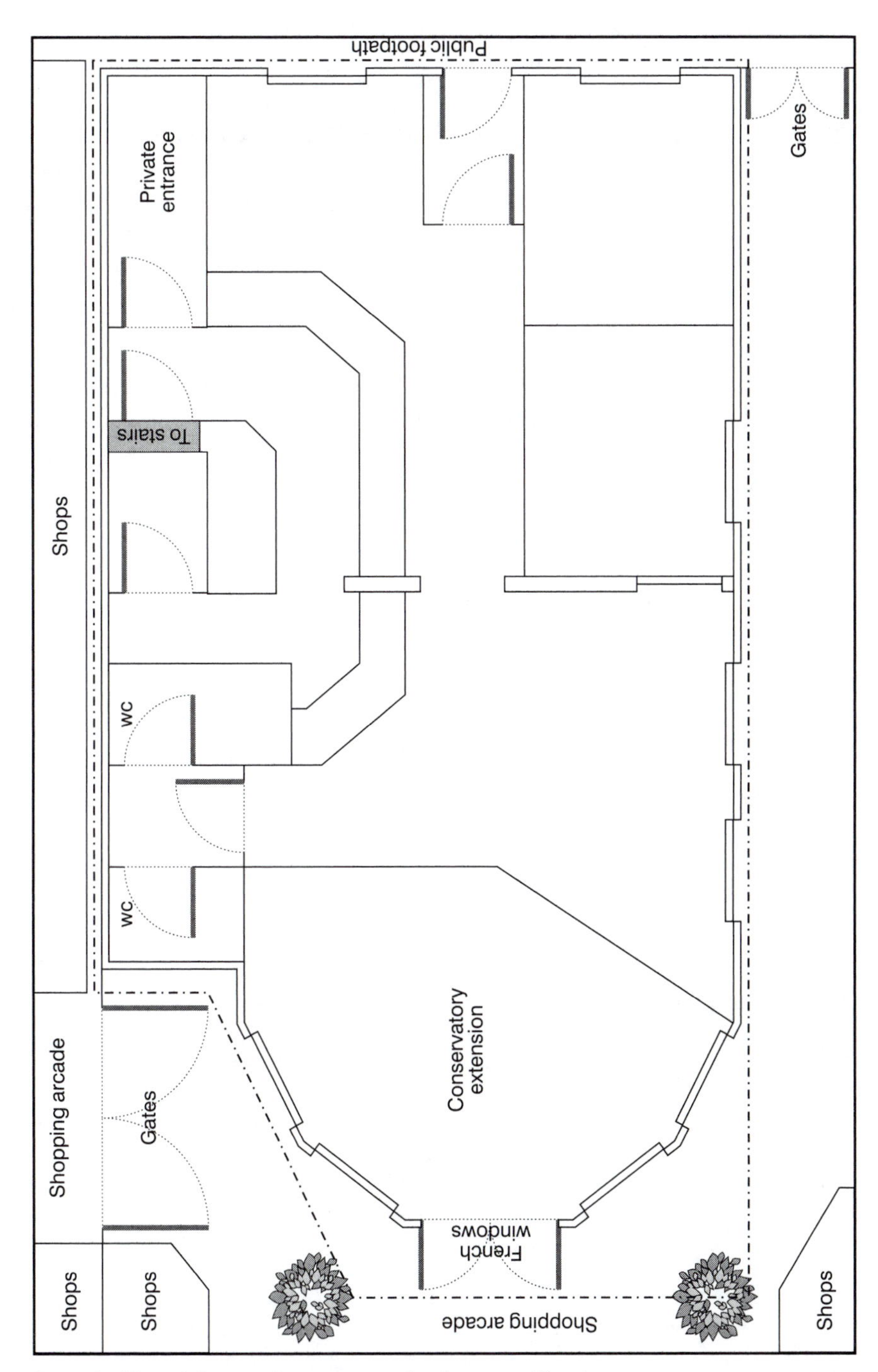

Sample Plan 4 Licensed premises or food service: Plan A

Keep a copy of this basic outline for future use in sections on health and fire, and a copy as Plan A in your storage file. The following Checklist 2.1. 'Site Plan A – Summary list' is a useful prompt list to complete to ensure you have included everything.

Checklist 2.1 Site plan A – Summary list of site features

	Noted (tick)	Query? (tick)	Action needed	Not applicable
External boundary				
Access routes for vehicles (roads, tracks, etc.)				
Access routes for pedestrians (footpaths, etc.)				
Position of external walls of main building				
Position of sheds, outbuildings, etc.				
Exterior doors, windows identified				
Car park areas				
Delivery areas				

2.2 External features

Plan B – External features

We shall add to the basic outline now, and you can either add more detail directly to Plan A, or use a tracing paper overlay.

The easiest approach is to think about first impressions of a visitor to the site who is unfamiliar with the layout, whether they are a customer, sales representative, worker or other visitor. Consider the following questions and add details to the plan as you go along. Make a note of anything that you think might need more attention, or you are not sure if it is adequate for your particular situation, to follow up later. If you put a small '?' at the point on the plan where action is needed, it can be blanked out once dealt with – see Sample Plans 1–4; Plan B.

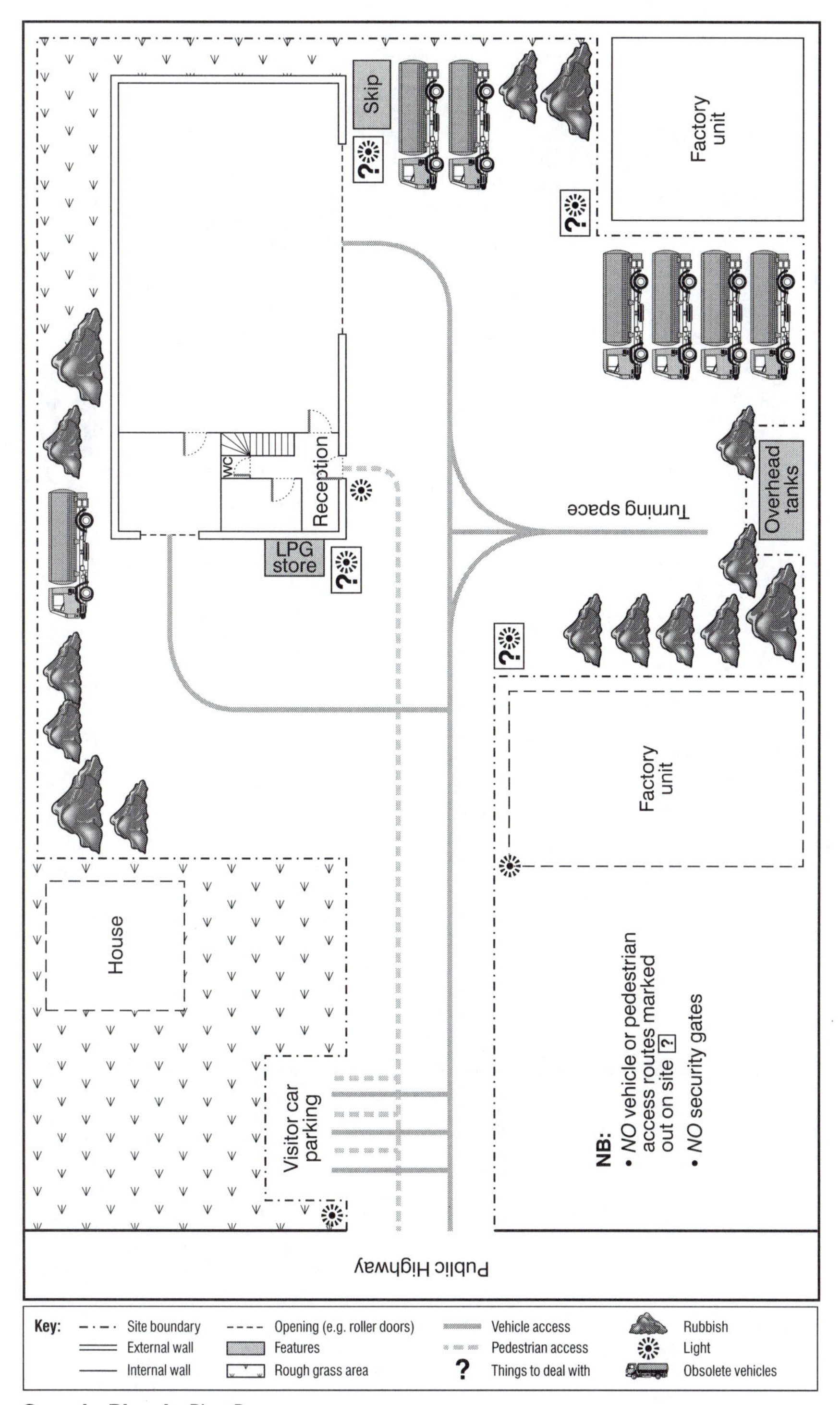

Sample Plan 1: Plan B

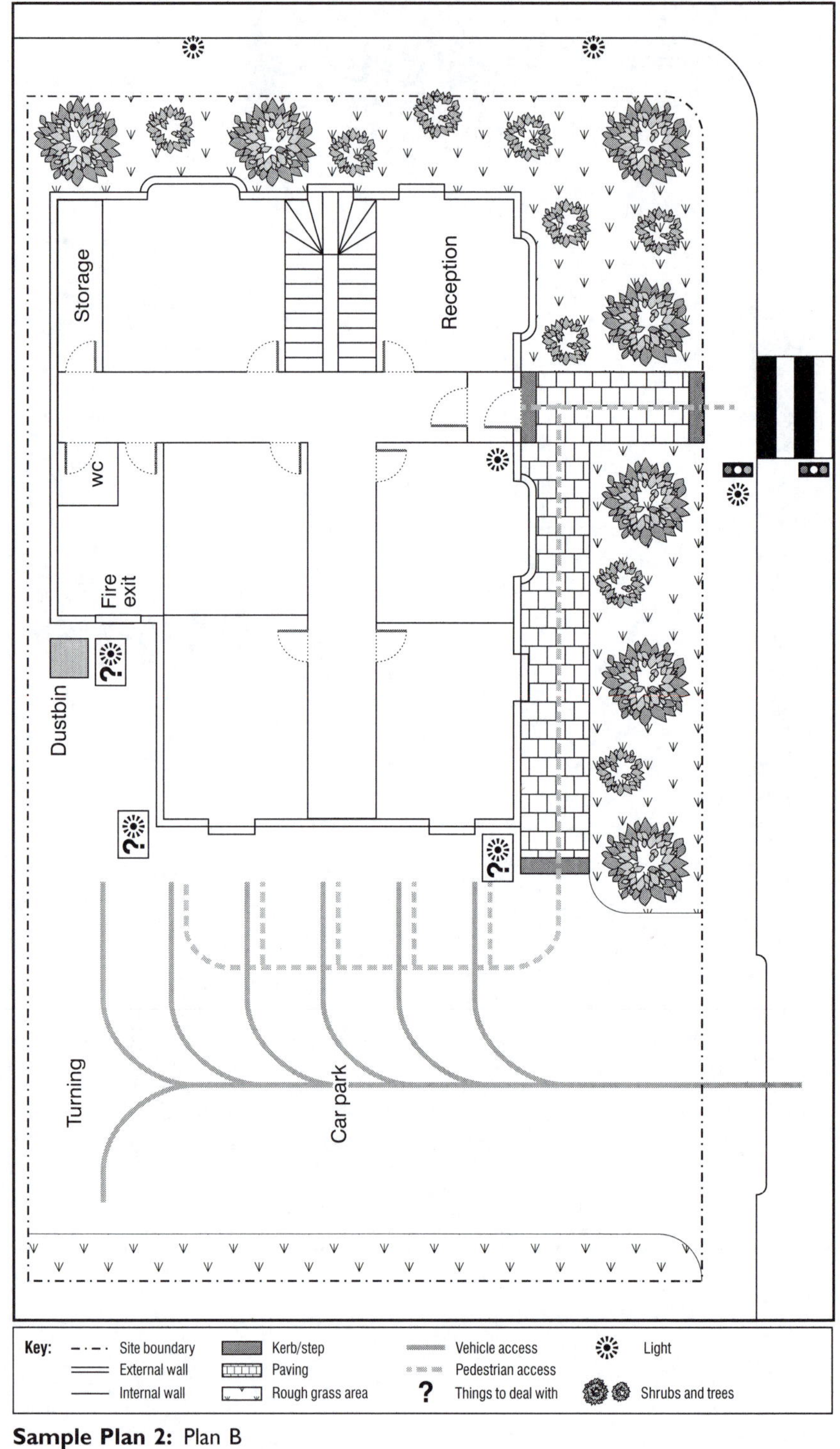

Sample Plan 2: Plan B

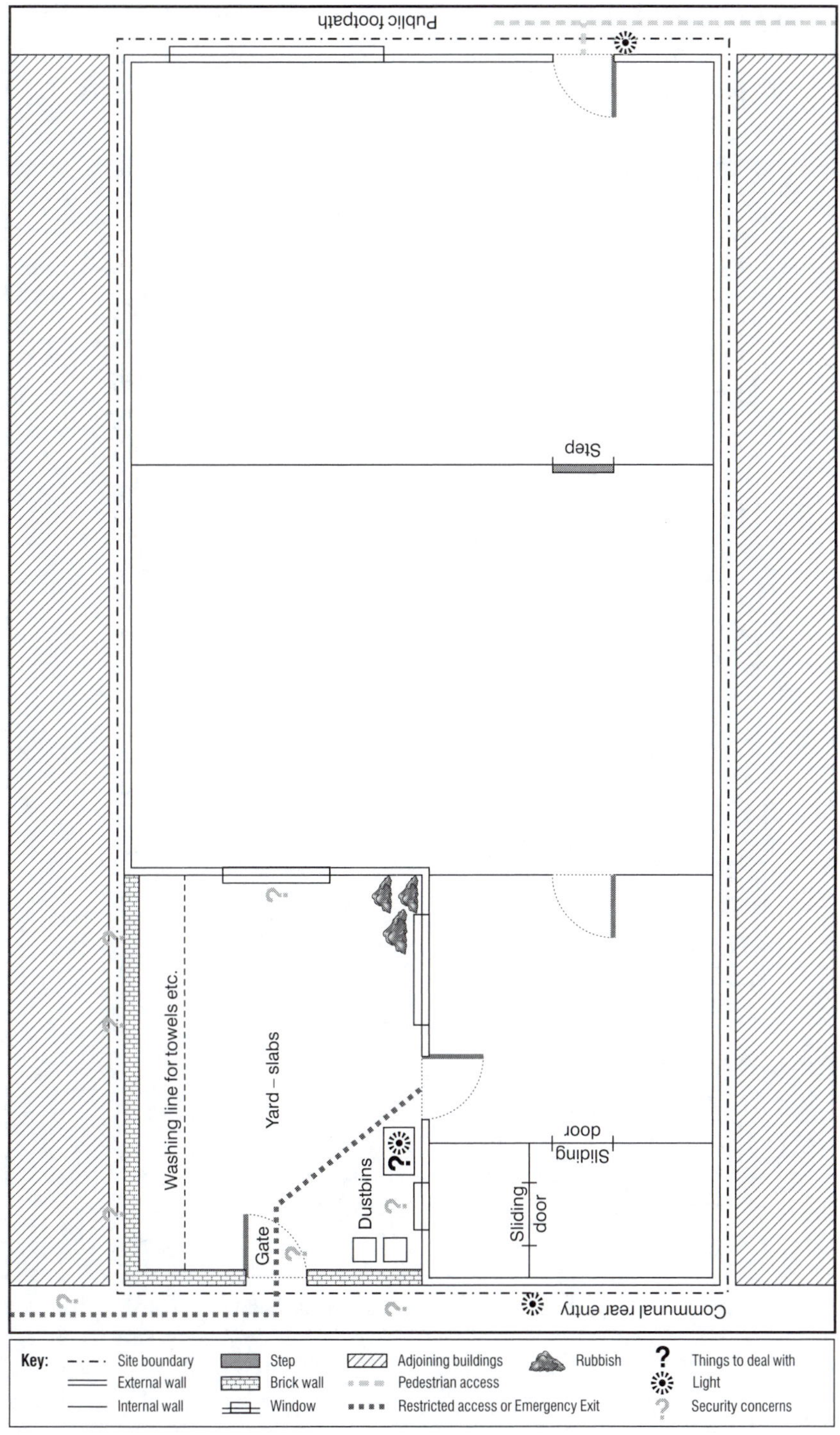

Sample Plan 3: Plan B

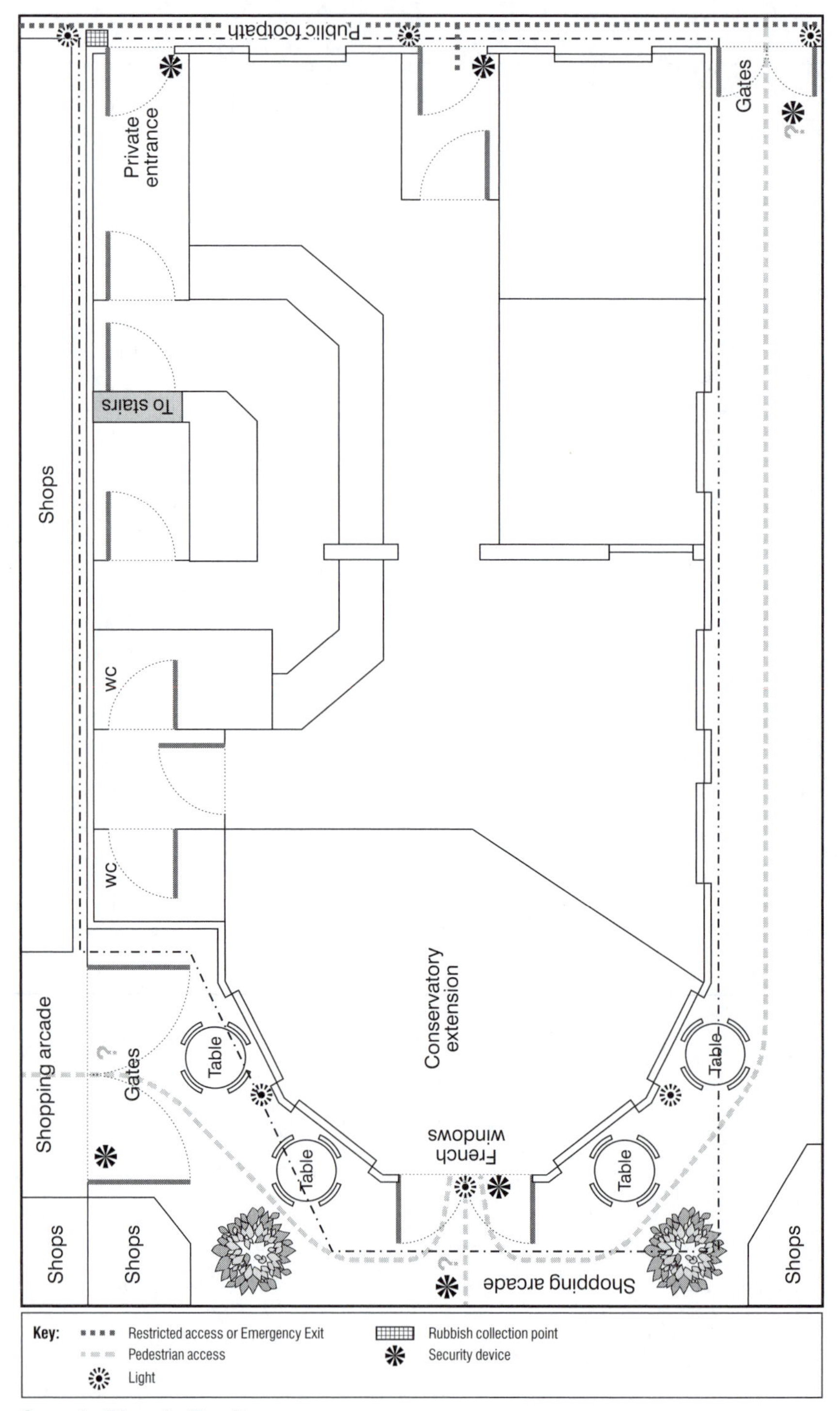

Sample Plan 4: Plan B

Questions and areas to look at and add to Plan B

Also see Checklist 2.2: Exterior Plan B – Summary list (p. 22).

All visitors

- Is entry to the site clearly marked for both vehicles and pedestrians? Is the reception area or main entrance to the building clearly identified, so that people are not wandering around looking for the way in? It also needs to be clear which entrances are NOT for visitors, for example direct access to workshop areas via open external doors. Note where signs are positioned, and whether they are clearly visible.
- Where are lights positioned? Are they adequate? Working? Sufficient for the area being lit? You should also note where there are dark spots or areas of deep shadow. Note where lights are positioned on buildings and in other areas.
- Is it necessary to separate pedestrian from vehicle routes, and if so has it been done? This may just involve painted lines on the ground, but could also warrant barriers of some kind in particularly busy areas or blind spots. This should become apparent as access routes are drawn on the plan, and crossing points identified.

Vehicles

- Are parking and turning areas clearly marked and kept free of obstacles?
- Is it clear where short-term stops can be made, for example to deliver mail or small supplies? If not, would it improve safety and security if a designated area was identified – where could you put this on your plan?
- Identify routes usually taken by fork-lift trucks, tractors or similar vehicles on site during normal working, and make sure you refer to 'Further Guidance' if such vehicles are used regularly.
- Surface of routes – while there may not be a particular specification for the surface of land around your premises, do identify any areas that are very poor and thus present a significant hazard to vehicles or pedestrians, such as vehicles overturning or people tripping.

Pedestrians

- Are there kerb edges or steps that people need to negotiate, and if so are they well maintained and secure? Identify any badly broken sections of paving, or step edges that would be more visible if painted white for instance.
- Also consider where lights are positioned in relation to these steps, and whether any changes or additions are needed.

Checklist 2.2 Exterior Plan B – Summary list

	Noted (tick)	**Query? (tick)**	**Action needed**	**Not applicable**
Signs for visitors				
Lights around site: lights on buildings				
Vehicle turning areas				
Parking areas including short-term parking				
Regular routes for fork-lift trucks and other vehicles				
Poor surfaces; kerbs and steps for pedestrians				
Waste skips; dustbins; areas where rubbish or waste materials deposited				
LPG, chemicals, flammable substances stores				
Storage tanks, for water, oil, or other liquids				
Obsolete machinery, equipment, vehicles, pallets, etc.				
Security: Perimeter gates and fences				
Security: Lights and CCTV				
Power lines and fuse boxes outside				
Roof details				

Figure 2.1 The use of fork-lift trucks presents many potential hazards. Note on the site plan where designated routes are or should be marked out clearly, points where visibility may be poor, lighting and surface conditions, and any obstructions.

Other external features to include on Plan B

- Waste skips and dustbins – where are they in relation to the building? Is this an appropriate distance; are they easily accessible by workers even when weather is poor? Is access restricted to unauthorized people, such as vandals or people dumping rubbish? Crucially, is the skip the right type for the waste being disposed of, and should you be separating out different types of waste for collection – check with the collection contractors.
- LPG and other flammable substance storage such as ink stores – are they stored tidily/appropriately according to suppliers' instructions/ in safe, secure purpose-built cages or buildings? Add them to your plan, and if you are not sure about any of these points then mark them FOR ACTION, and check with suppliers and other guidance.
- Storage tanks for water, oil, etc. – again, include them on the plan, noting whether they are appropriate, adequate, secure and well maintained. It is particularly important to make sure you know whether you have sole responsibility for such storage, if it is a joint responsibility with the landlord, or with other site users.
- Features that may now be considered permanent but which are, in fact, removable such as areas where rubbish or waste bits and pieces collect over time; obsolete vehicles, equipment, or machinery that may have been kept 'for spares' but which will never be used; damaged pallets or

Figure 2.2 Note where skips are located, especially how close they are to buildings and overhanging shrubbery as in this photograph. Are they easily accessible for use and collection, the right type for likely waste, clearly labelled, or left too long before being emptied?

shelving/racking that should either be sent back to the suppliers or disposed of in an environmentally friendly way.

- Apart from the fire hazards these features present, which we will look at more closely later, there are always the safety hazards of people tripping and being injured, rubbish piling up haphazardly with the possibility of falling, and of course poor visibility or obstructions for people driving on site.

Security

The sample business plans included here have very little in the way of security, with no restriction to access on the rural site – not even gates – and no security lights, window locks or control of people on site. Why security is included here is so that it is looked at in conjunction with health, safety and fire to ensure the different elements fit together. There are lots of specialists who can give you specific guidance on security measures you need to take, but you do need to be clear what you want to secure the premises against, for example, trespass/burglary/theft/vandalism/arson/or access to sensitive material.

Whatever the purpose of your security measures, some of the points you might consider include:

- perimeter access – walls, fences, gates;
- control of vehicles and people on site;
- Security lights, CCTV;
- locking devices on doors, windows, storage areas.

(Add details of these features to Plan B where appropriate, or include '?'.)

Other issues might be data protection and IT security; secure areas for chemical or pharmaceutical products; or indeed anything that is easily transportable with a reasonable street value.

Photographs of the premises, areas of concern that you need to deal with, or parts which you have controlled particularly well, will also be useful in the 'evidence' file, to demonstrate that you have actually started to look more closely at the workplace. It should also highlight aspects of the building structure itself that may present potential hazards. External power lines or fuse boxes may need special attention, either as a source of danger when accessed illegally, or indeed as overhanging cables across busy routes used by large vehicles.

Roofs are critical, whether you are in single or 2/3 storey buildings, the main problems being ease of access to burglars; adequate access to and maintenance of damaged areas; and how fragile the roof surface itself is (for example, old corrugated asbestos). We shall come back to some of these points later, but it is worth making a note of them now.

2.3 Internal features

Plan C – Main internal features

You can either add these details to Plan B if there is sufficient room, or make an additional tracing as Plan C. The sample plans show how just the outline of the building has been traced, and an internal floor plan for the ground floor has been produced. If people work for you on more than one floor, you will need an outline for each unless the layout and activities are virtually identical. However, when we look more closely at fire risk assessments, there should ideally be a separate plan for each.

Include all internal walls, passageways, doors and stairs. Also include fixed storage racking/shelving and workbenches, but use a different symbol for each. If portable shelving is used to divide working areas, then include that too. As you can see from the examples, overhead power supplies and large pieces of machinery have been identified in the manufacturing firm, and computer terminals in the office environment. It is useful to identify toilet, washing and eating areas, as well as the main large pieces of furniture, equipment or machinery.

It is also worth noting at this stage where fire extinguishers, smoke alarms, etc. are located, with a red '?' if they are missing or not working. As we shall be looking at individual work areas in more detail later on, at this stage we just need to label the main work areas on the Plan C.

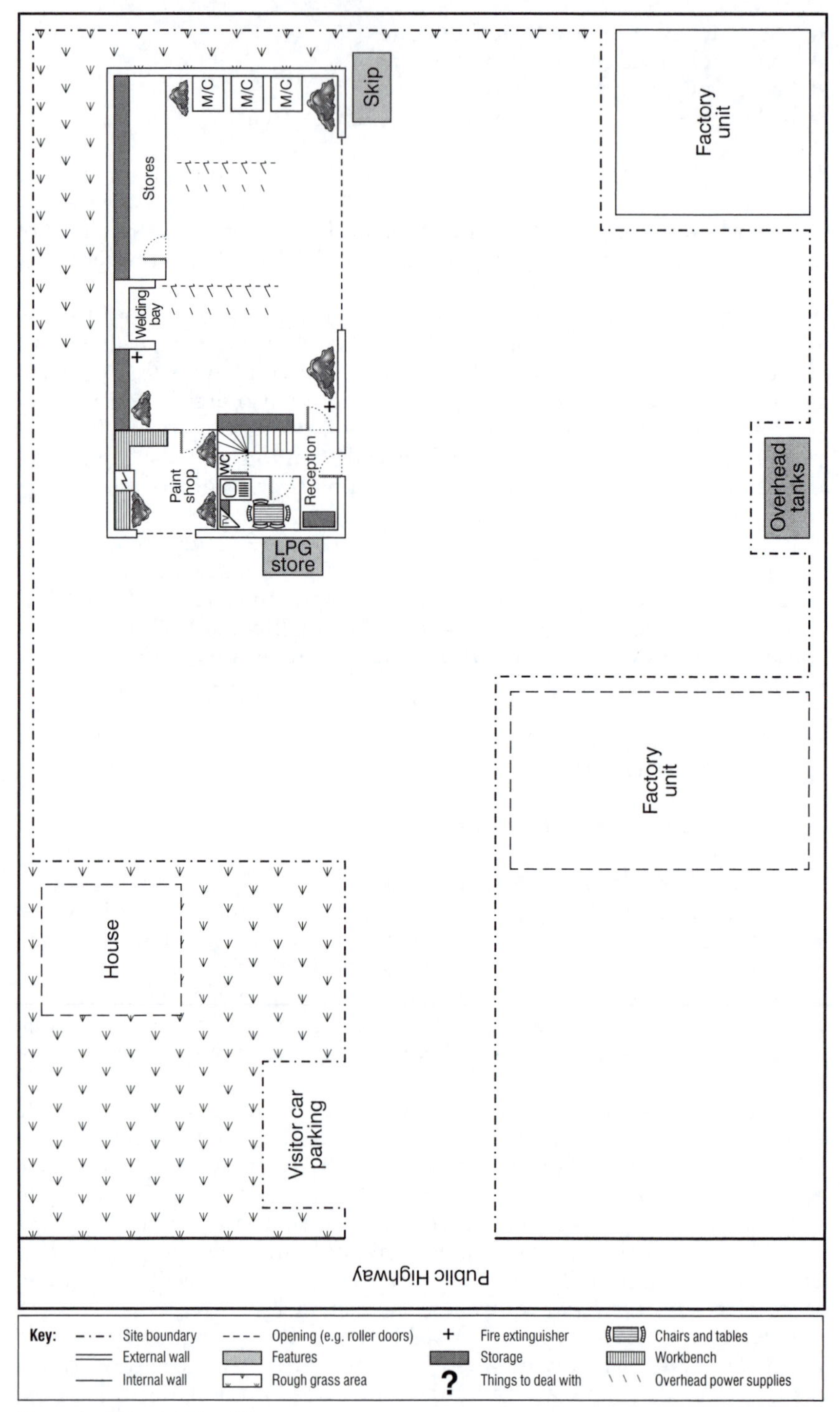

Sample Plan 1: Plan C

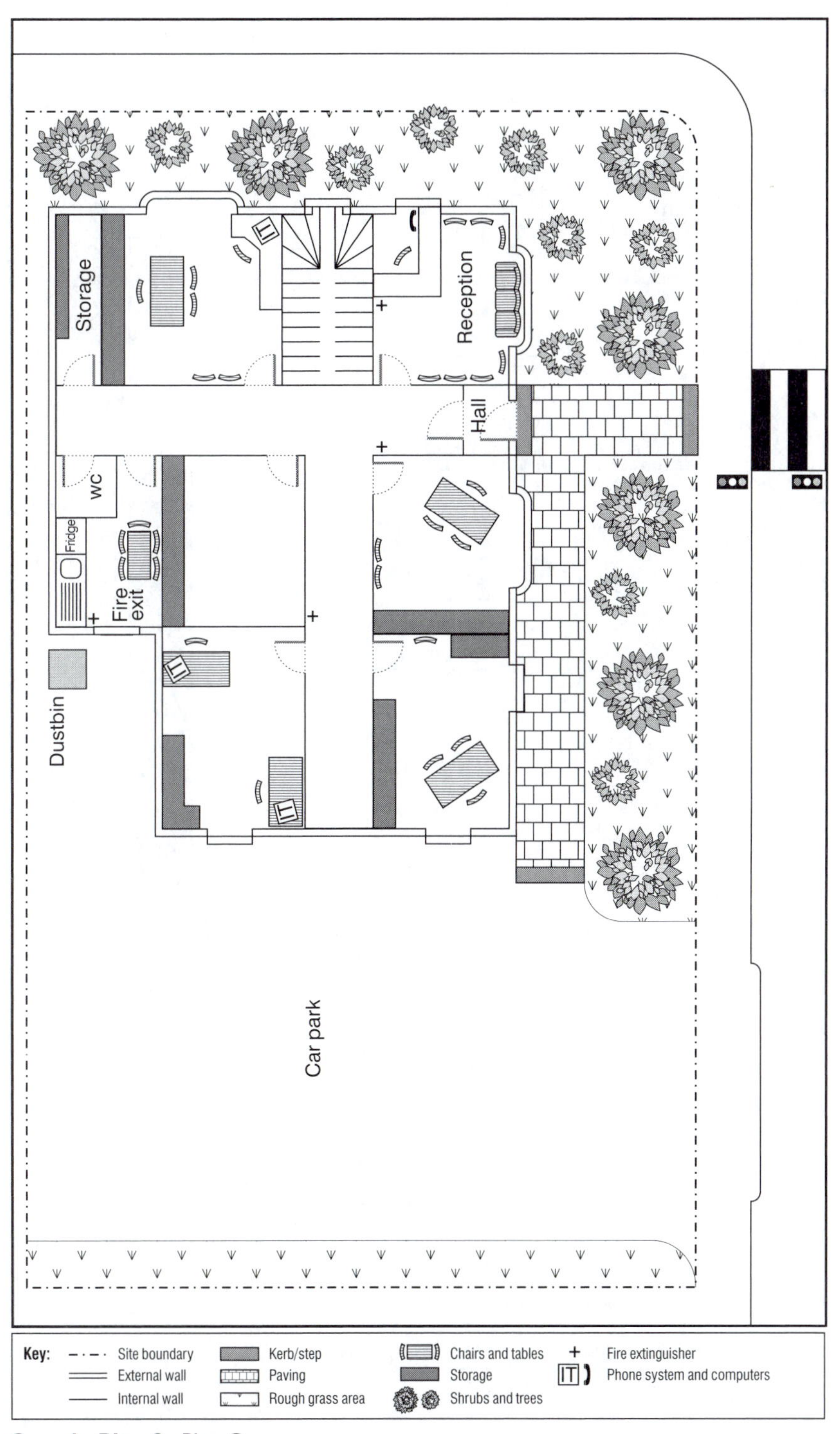

Sample Plan 2: Plan C

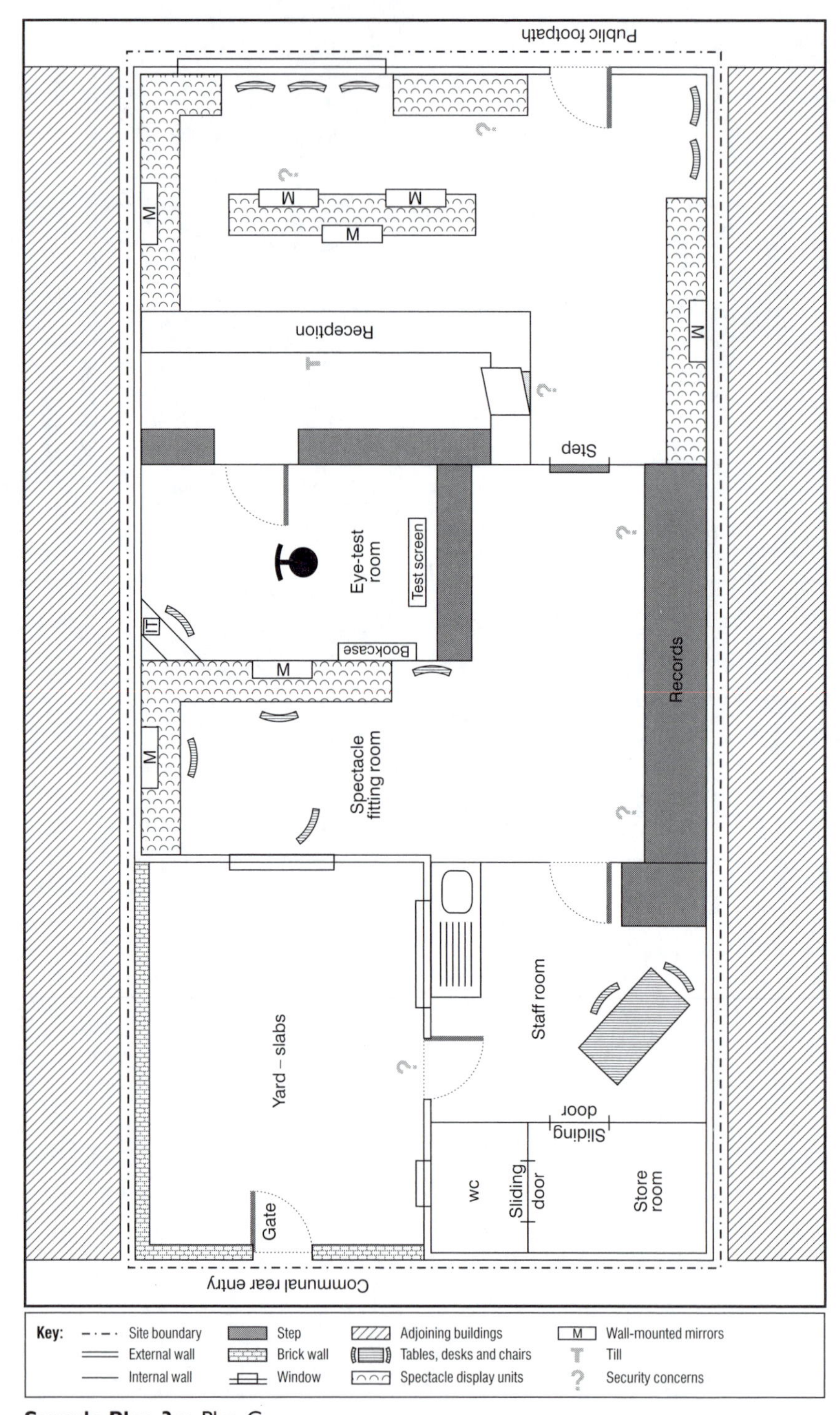

Sample Plan 3a: Plan C

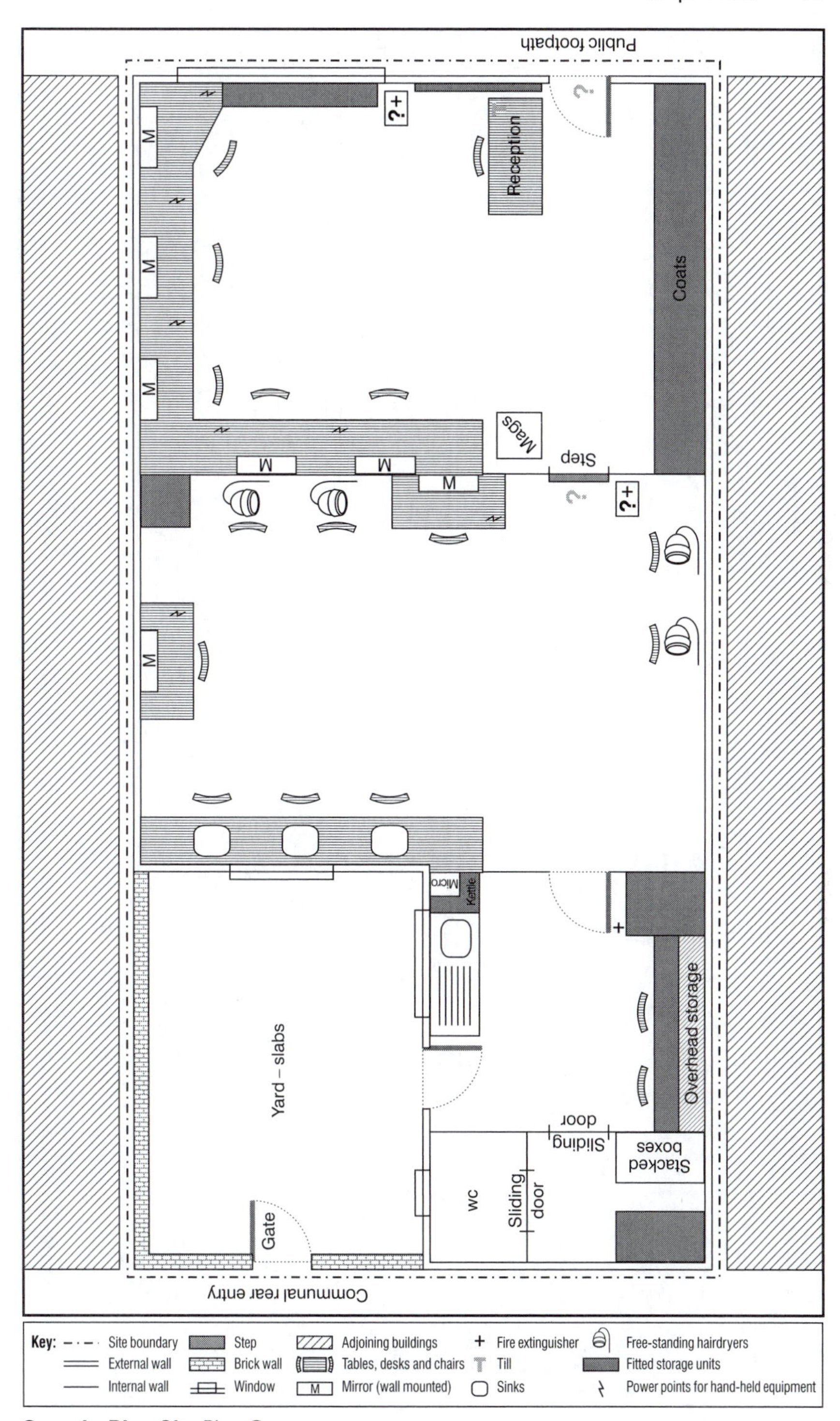

Sample Plan 3b: Plan C

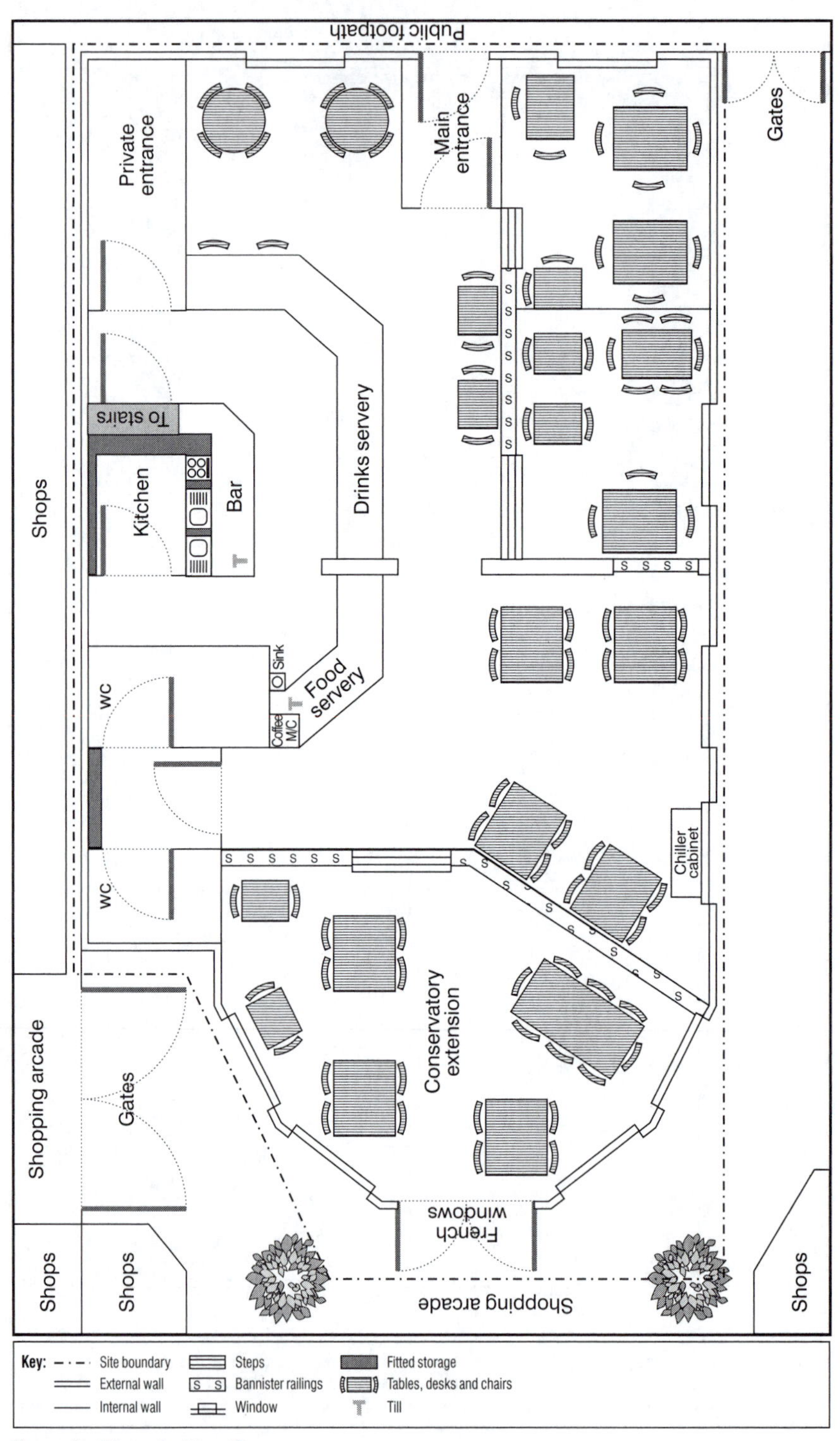

Sample Plan 4: Plan C

Checklist 2.3 Internal Plan C – Summary list

	Noted (tick)	Query? (tick)	Action needed	Not applicable
Plan for each floor of premises				
Internal walls				
Stairs				
Doors, including sliding or folding doors				
Fixed storage shelves or racks				
Portable storage				
Fixed workbenches or surfaces				
Overhead power supplies				
Large pieces of equipment or machinery				
Large pieces of furniture				
Computers or IT equipment				
Toilet and washing facilities				
Eating areas, kitchens, vending machines				
Fire extinguishers and smoke alarms				

A photograph of each work area or section would be invaluable here, with details noted alongside of:

- main activities carried out;
- where machinery and equipment is used and stored;
- the usual position of desks, tables, chairs, computers, shelving, tills and so on.

Several examples are included in the sample plans for a variety of typical small businesses, so you may be able to use one of these as a skeleton plan for your own firms. Checklist 2.3 'Internal Plan C – Summary list' (p. 31) is included here to act as a prompt list.

Chapter 3

Processes and movement of goods through the business

Where does an order come in? What happens when you receive an order? In some businesses it may be that it starts before this point, but for now we will concentrate on the fact that you have the order, and you know you have to fulfil it. So, what happens to it? If we take the example of tanker repairs and the plan we have, it may be that an outside driver drives on to the site, bringing in the tanker that has to be repaired. He parks it somewhere, possibly where he is told to, possibly where he wants to, or maybe where there is a space left. And then he walks into the building, probably into Reception, and waits for someone to meet him. Does he have to wait a long time, will he be wandering about into the actual factory space itself, or is there a designated area where he can wait until someone sees him?

He has now delivered the vehicle. Where does the vehicle go next? Someone has to drive it from where it is parked and take it into the factory. It stays there for a while, production processes take place, then when it is completed, there may be testing arrangements to be made, and it has to go somewhere else to be tested. But otherwise it may just go out into the yard to wait for collection. At what stage did the driver leave, or did he just wait? How does he get back if he has to drop off the vehicle and leave it? Is he the responsibility of his company or the one carrying out the repairs at this stage? These are just basic points that you will already have thought about, but you need to be clear about procedures to the point where someone comes to collect it and drive it off site.

3.1 Arriving on site

When people come onto your site, what sort of security measures are in place? For instance, do they have to sign in when they arrive (which is also an important point when we look at fire hazards and risks later on)? Are there gates at the perimeter to act as a barrier to unauthorized personnel, and if not should there be? This may not be a realistic option, due to lots of factors such as location, ownership of land or premises, or even accepted practice, but it is something that may warrant further thought if only to dismiss it. How do you monitor movement of visitors

Figure 3.1 Identify major vehicle and pedestrian routes on the site plan, particularly where these overlap. Consider whether pedestrian routes should be marked out, and whether barriers or warning signs are needed. Also note whether loading areas are in the most appropriate place for safety.

on site, especially if there are restricted areas that are particularly dangerous or sensitive? If you look at your Plans A and B alongside the sample plans – are there any safety issues here about how people and vehicles move about the site?

It might, on the other hand, be a service that you provide and the customer comes in to see you. Where do they come in and where do they wait to see you? What sort of procedure have you got for booking in and out? How do you know that people have arrived, where they are waiting, the route by which they will leave after making appointments, etc. Again, this should not be too complicated, but quite easy to follow through on your own plan. Of course, there may also be particular time scales for each of these procedures, or it may be flexible and depend on the job. If possible, then do identify these details on Checklist 3.1: Movement of goods.

Virtually all businesses will receive deliveries of goods or materials, if only the mail. It may be an extremely minor element of your particular business, just requiring a couple of sentences to acknowledge that you have thought about it. On the other hand, there may be quite a wide range of people coming to your premises to collect or deliver items, so the following points may need to be considered as they all represent potential hazards of some kind.

Figure 3.2 If the reception area for clients on site is attractive, warm and comfortable, as in this photographer's Studio, they are more likely to stay there rather than wandering into hazardous or unauthorized areas. Adequate seating and lighting is important, but also make sure exits and escape routes are clear.

- People often leave vehicles unattended, sometimes with doors open.
- Who is responsible for loading or unloading goods? Do your staff sometimes help?
- Are your staff required to move other people's vehicles sometimes, and indeed should they be?
- Do you know (or need to know) the extent of activities the driver of the vehicle is responsible for, especially if they are employed by another firm – what have they been told they can or cannot do.
- Is there special equipment available to transport items from the point of delivery, for instance can they be moved by hand or do they need a trolley or truck?

There is a much greater reliance now on the use of standard size and shape of pallets, which can certainly help when larger quantities of goods are involved. However, for most small firms this is not generally the case, and smaller quantities are often delivered in containers of all shapes and sizes. In addition, having identified all these points about delivery, there are further security questions such as how do you check the contents of delivery? Are adequate records kept, and are they secure? Where are deliveries stacked on site at time of delivery, and is there a designated area where they should be? There may also be questions about whether there are hazards associated with where or how goods are stacked.

3.2 Storage

Supplies – what happens when you have to order supplies to be able to carry out the job. Where are orders processed and who decides what needs to be ordered and when? In addition, you need to identify where goods arrive, where they are stored, and how they get to where they have to be used. Adequate storage space and facilities may be an issue for you, such as boxes of stationery stored on the ground floor when delivered, but actually used upstairs where there is little space available. This might mean that someone has to regularly walk up and down to collect items, and while this might not in itself present any great problems, the more journeys up and down stairs, the more likelihood of them putting themselves at risk of tripping or falling. As you have already identified the storage areas on your plans, it should be quite straightforward to identify what sort of stacking systems are in place, where there are secure storage areas, and what control you have over access to and distribution of supplies.

Figure 3.3 Specially designated storage units are commonly available for flammable or toxic substances, such as oils, solvents and cleaning chemicals. Units should be clearly labelled, well maintained and secure, and placed at appropriate distances from buildings (see suppliers' instructions). Look out for empty containers carelessly stacked or discarded causing health, safety and fire hazards.

There are additional issues related to 'hazardous substances', which includes solvents, inks, bleach-based products, poisonous substances and many cleaning agents, even if you only use small amounts. However, there may be more comprehensive requirements associated with some hazardous substances, such as how they are stored, where they are stored, whether they should be kept in special containers, what they are stored next to, and whether they should be kept at certain temperatures. We are not talking about food or perishable items here, although these will also be an issue for some firms – whether it is to feed staff, a saleable commodity, or will be processed in some way. Storage is important not least because things are often stored anywhere for convenience, far too high up for people to reach comfortably, on very rickety shelves, and in a more awkward position than they need to be, so increasing likelihood of personal injury significantly.

3.3 Reception area

Again, this is an area often overlooked, often an afterthought, but also potentially an area of concern regarding health, safety and security in some cases. While the reception area may be a crucial part of the customer contact for many small firms, for others it is not appropriate to have a receptionist in attendance all of the time. Questions to consider here include whether there is a procedure for signing in and out of the premises, and whether visitors are actually given information about the site before they enter – we shall come back to this question in the procedures section. If this is something you need to think about further, then put a '?' on this part of the plan.

Now you have people on site, consider where they sit, whether any refreshments are available, and is access to some parts of the site restricted to them? For example, drivers who deliver vehicles for repair may have to wait for an hour. They can sit in a designated area, but it is very dirty and scruffy, with chairs with unsuitable stuffing (fire hazard) – in fact, very depressing! So they become bored and drift around, wandering into the actual workshop to see what is happening, and talk to people carrying out repairs. Clearly this is unsatisfactory, and despite notices being displayed around the area telling them it is not acceptable, no-one actually enforces it, so again it may just be a case of identifying where people are at any given time, and the purpose of the reception area. Various options are available to deal with these issues, such as the use of security badges, visitors escorted around the site, using CCTV, or perhaps numbered access locks on doors for staff, but these can be considered later.

3.4 Progress through the firm

Now we have gone through some of these points in detail, you should be able to work through the different work areas or sections on your plans to produce some sort of picture of the movement through the business.

Put it onto Plan B or Plan C, or an extra one if you want to. It may be useful to use another plan, as we still have other details to add. The following are just a few pointers to give you further ideas as you complete the tasks.

- Materials coming out of stores and going on to the first process stage – what do people do with them, how are they transferred to next stage? Are they easily moved as on a conveyor belt; does someone turn around and pass it on to the next person; does someone have to carry it?
- Having completed all the tasks, how does it get to the customer – is it stored while waiting for collection? Processes involved in packaging are often forgotten as part of the process, yet quite complex or hazardous pieces of equipment used, including staple guns and cling-wrap machines. Once the item leaves the site, where does your responsibility for it end?
- In a service industry, the customer may be the 'Work in progress' in that they move from one work area to another, so using the same principles, where do they go at each stage? In the same way, when you go to another business' site or the customer's home, you or your staff need to be alert to the same sort of questions.

Checklist 3.1 Movement of goods through the business

Stage of progress through firm	**Area on site Plans B or C**	**What happens at this stage?**	**Who deals with it?**	**Specific security measures in place**
Stage 1: Arrival				
Stage 2: Process				
Stage 3: Process				
Stage 4: Completion or finishing				
Stage 5: Delivery				

3.5 Procedures

In this part, you need to look at what procedures are already in place, and what evidence you have to show that relevant people know what these are. We shall eventually be able to see if there are any gaps in provision from a health and safety point of view, but shall start from the point of 'what written procedures already exist in the firm?' Having identified the different processes involved, you should now be able to return to the plans to see

- if there are procedures already in place;
- if they are written down;
- when they were put together;
- do they need to be amended (which means you will have to look at them to check!).

Now is an ideal opportunity to review these procedures with the people involved to see whether they are workable, relevant, appropriate and, of course, safe. So go through the different sections to check whether this is the case.

To summarize

Check procedures for arrival and departure of visitors; movement of vehicles on site; how goods in and goods out are monitored; how people pass on goods to the next stage in the process. If you have a formal quality system in place, it should already form the basis of this part of the task. If there are large or complex pieces of machinery in use, there may also be formal 'safe system of work', or specified 'permit to work' procedures in place, or possibly guidelines from manufacturers or suppliers of goods, equipment, machinery. So collect all these documents together and keep in the evidence file, or at least be able to access them readily when needed.

For hazardous substances in particular you should already have a system in place for keeping Hazard Data Sheets. If you have not heard of Hazard Data Sheets (HDS), then basically any chemicals that have some form of 'danger' sign on the container, are likely to have a HDS that the manufacturer has produced. Although there is as yet no one specified format for such sheets, they broadly follow the same pattern in that they tell you exactly what the hazard is, correct storage and handling procedures and protection for users, and relevant treatment in the event of mishandling. (See References for more details of the Control of Substances Hazardous to Health COSHH regulations.) These details need to be collated, along with any relevant manuals, but clearly you do need to check that everything identified in the last section is included.

Having identified the main processes that take place, we can now move on to the next section and look at the way people carry out the activities, and what procedures are in place to complete each task.

3.6 Activities in each area

This is an important stage, based on the flow of work you have already identified, and should include all areas on site. What happens in each section? It is much easier to involve other people who carry out in the tasks at this stage, to clearly identify who does what where. We are not really thinking about 'hazards' specifically, but rather who carries out specific activities in different parts of the firm – even if there are only three people working there!

Make a note of what happens at each of the stop-off points and work areas on your progress plan, then look closely at the equipment or machinery that is used in each place, perhaps at different times of the working day. What do people actually do at each stage? It would be extremely useful to take photographs of each major work stage, both as a reminder and to refer back to at later stages, for example, waiting areas or the paint shop.

An example of completed Activity Checklist 3.2 is shown on page 41.

It may appear more complicated when recorded in this way than it is in practice, but it is a valuable exercise that can illustrate some significant issues not previously considered. It is important to include storage areas as well, especially those that are regularly visited by staff, with notes about how supplies are transported in and out of the area. As we started

Figure 3.4 What are the main activities carried out in each work area? For example, in this cooked food preparation area in a butchers, there are washing facilities for hands, utensils and vegetables, and the large oven is near to working surfaces. Also consider whether preparation and storage areas are adequate (note large dishes stored on top of oven), equipment is suitable for the tasks, and traffic flows do not add extra problems for people working there.

from the Site Plan and external features, do not forget to note the activities that take place outside the building, and the subsequent use of other equipment or machinery.

Area	Activities that take place	Equipment used	Number of people involved
Reception	Take deliveries Meet visitors and sign in Make tea and coffee Use telephone	VDU Telephone and switchboard Kettle	1
Waiting and rest area	Tea, coffee lunches Customers can watch TV	Kettle and microwave	4 maximum at any one time
Main shopfloor area	Removing engine covers Checking vehicle components and structure Fit replacement parts	Hoist; compressed air; overhead drilling and riveting tools; portable electrical equipment; LPG Trolleys and trucks	3
Welding bay and paint shop	Mig/gas welding Spray painting bodywork Spot repairs by hand Spray cleaning vehicles when necessary	Welding equipment Compressed air lines Hand held tools	2
Office	Administration – use of computers, copying machines; keeping records Filing; talking to customers by phone	Computers/VDUs Printer Photocopiers Telephone	1
Stores	Storing office supplies Storing chemical substances for shop floor Tools and components	Small trolley Range of shelving and racking	1 occasionally

Checklist 3.2 Activities in each area

Area	Activities that take place	Equipment used	Number of people involved

Part B

Controlling Safety Risks

Chapter 4

Safety and security hazards

Having drawn a fairly comprehensive picture of how work progresses through the firm, we can start to look at some of the more obvious hazards that may exist in each of these areas. In this case, we are defining hazard as 'something that could potentially cause harm, injury, or damage to people and property', although the likelihood of it doing so may be quite small. We have included damage to property as well here, as sometimes this could itself lead to harm to someone at a later date and indeed we want to make sure that your insurers are happy that things are under control. Later chapters will consider health and fire risks in more detail, but for now we shall concentrate on safety and security issues in each of the work areas we have identified.

There are lots of different ways you can define a hazard, and many different ways that you can organize your hazard checking lists. In this case, we shall start with a list of headings that reflect the types of machinery, equipment or activities that you are likely to carry out in your business. Use Checklist 4.1 Hazards on site at the end of this chapter to record your findings.

4.1 Vehicles

The main hazards relate to the possibility of being run over by the vehicle as it moving about, and it is particularly hazardous when a vehicle is reversing, especially when the driver's view in the mirror is restricted. You may have seen signs on the back of lorries as you are driving along, pointing out that if you cannot see the driver reflected in the side mirror, then he cannot see you. It is sometimes easy to forget that the driver does not see you in the same way that you can see the vehicle.

Other hazards associated with vehicles are being crushed, either behind or at the side of the vehicle. If the space in which the vehicle can turn or manoeuvre is limited, especially on very small or crowded sites, there is a real danger that someone can be crushed alongside the vehicle and a wall or other obstacle. Also note potential problems of vehicles overturning where the ground is not firm or is full of holes.

4.2 Machinery

There are many issues around moving parts in machinery, and clearly your particular business will use different types of machinery and equipment, but some of the main things that you are likely to find include:

- Workers' clothing being trapped by the machine, and people's hair or jewellery can be caught.
- Being hit by moving parts of the machine, and of course don't forget materials that can be thrown out by the machines becoming very fast projectiles with the potential to inflict some real harm or damage to workers or passers by.

Figure 4.1 Machinery safety – identify the main items of machinery and consider the potential hazards, such as secure fastening; electrical parts regularly checked; proper guards fitted and used correctly; good local lighting and exhaust system. Note storage above and behind the machine in this carpentry workshop, so safe working procedures need to be followed to avoid reaching across the machine when working.

- Bands or cords that move quickly have the potential to break and snake out, causing a whipping type of action which clearly can cause severe injury.
- Abrasive and rough surfaces may represent a hazard, as do machines that spin very quickly such as centrifuges or dryers.
- Don't forget to look closely at guards and handrails on machinery to identify potential hazards when in use.

4.3 Sharp tools and objects

These can form part of many different types of machines, and could be used for chopping, cutting, mincing, and shredding. Clearly any machine that is capable of cutting or chopping materials can cause significant damage to human fingers or other parts of the body.

The use of knives and blades for cutting, whether it is for food preparation or for opening storage boxes for instance, are potentially very hazardous items to note. Even in situations that are considered to be fairly safe, such as office environments, include some hazards such as the use of guillotines for cutting paper, and even the sharp edges of paper itself.

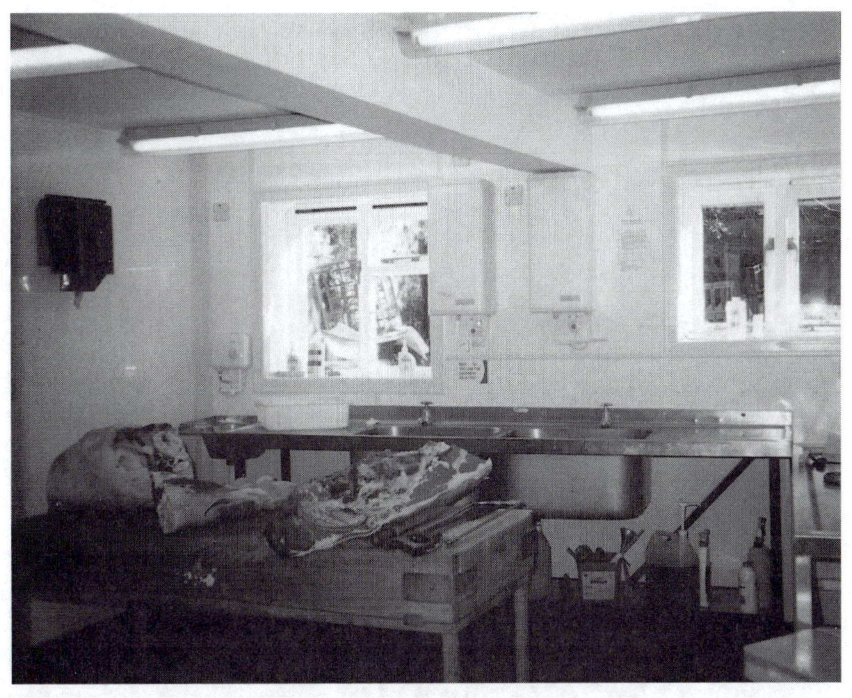

Figure 4.2 Consider where and how sharp knives or tools are used by staff, where they are cleaned after use, and whether they are kept safely and securely when not in use. In situations such as this butchery section, the correct personal protective equipment (PPE) must be available and worn, and emergency first aid provision kept nearby.

These may not be significant hazards, but they should still be recognized by you. Other sharp instruments may include hand drills, and sometimes quite small hand held pieces of equipment.

4.4 Heat

Burns and scalds hazards are generally quite easily recognized by people working in situations where heated objects or materials are involved. But it is also important to note that apart from working in kitchens, there can be potentially hazardous situations when using glass washing machines in pubs for instance, and even when using a photocopier if the paper becomes jammed and the machine has to be opened. In situations where people are using welding machines, then sparks flying and the heat itself clearly represent a hazard to the person involved. Also remember that in extreme cold situations there could potentially be burns and frostbite damage as well, so try to be as broad-ranging as you can be when looking out for hazards.

4.5 Electricity

There are many potential hazards associated with the use of electricity in the workplace. Not just large electrical pieces of equipment or the most obvious ones, but sometimes very small hand-held pieces of equipment

Figure 4.3 Consider electrical hazards in all work areas, including offices and storage areas, particularly old or worn cables, overloaded sockets, trailing wires and cables, especially in areas of through traffic.

can be just as hazardous to people. Portable pieces of equipment need to be regularly checked to make sure that they are in good working order, but at this stage we just want to identify where such equipment is used in your workplace and where potential hazards exist.

Make sure you note overloaded or damaged socket points, and trailing cables across busy pedestrian traffic routes. It's also important that you consider the rest areas and places where people make cups of tea and coffee, as you can often find kettles placed on sink draining boards, which is clearly not a good idea! Later on we shall consider how you can safeguard people from such hazards, but at this stage we are just identifying them.

4.6 Working at heights

Where do you actually use ladders, steps, or scaffolding to carry out work in your business? It might be for regular activities such as in a warehouse for instance, or perhaps just for certain activities that do not happen very often, but nevertheless require the use of steps or ladders. Hazards associated with working at heights are not just about the person falling, but include dropping things from a height causing injury to other people, plus issues about working on fragile surfaces such as glass or asbestos roofs.

4.7 Confined spaces

This is usually associated with workers who have to get inside tank bodies, for instance, or underground workings. However, it also applies to staff working in cellars in pubs and other licensed premises, and in fact anywhere that has limited access, movement and breathing facilities. The hazards you may identify, therefore, are more likely to be related to situations where the person has to sit or lie in cramped areas, with very little room to manoeuvre themselves or equipment, and perhaps where they generally only work for very short periods at a time. It is vital that you do identify situations where people are likely to be using welding or blow-torch equipment inside vehicles, vessels or other containers.

4.8 Use of compressed air or LPG

Both of these represent hazardous circumstances, so make a note of where either of these properties are used, plus other instances where substances are contained under pressure.

4.9 Slips, trips and falls

As the major cause of most workplace accidents, you should look carefully for areas where these hazards are present, and particularly at standards of general housekeeping. This will include such things as:

Figure 4.4 In many older premises, the stairs may be poorly-lit, uneven, narrow, steep and winding causing very real tripping hazards. Ensure this is not made worse by obstructing stairs with old containers or materials, additional tripping hazards, or using temporary step coverings such as the strips of carpet shown here.

- shiny, smooth floor surfaces;
- spillages are likely, of water/oil/grease/powders or dusts;
- waste materials and scrap lying around the floor;
- poor visibility where doors or windows open;
- changes of floor level, steps and stairs;
- obstructions in passageways regularly used by people, in fire exit routes and on stairs;
- loose, rough, or worn edges of carpet and other floor coverings;
- cables or air lines.

4.10 Lifting and carrying

Both manually and mechanically – noting potential problems related to the size and shape of loads; the weight in relation to other elements (for example, the weight combined with the action of silver-service

Figure 4.5 CD: Work in agriculture or horticulture often involves repetitive twisting movements, as when picking the grapes at this vineyard. By sitting at the correct height for comfortable reach, the potential harm is reduced.

waitresses/waiters as they support it on outstretched arm); people having to twist or move awkwardly once they have lifted the object. It is important to look at areas of storage, the height of shelving, and likelihood of sharp edges in packaging or objects carried, and of course lifting people or animals.

4.11 Repetitive strain injuries

Associated with regularly repeating the same movement, so putting a strain on certain joints in the body. This can be associated with the use of computer keyboards, and production areas where workers twist in their seat to complete different parts of the task. There are also hazards associated with using vibrating machines for any length of time, such as some drilling machines.

4.12 Chemicals

We shall look at these in more depth in the Health section, but need to identify here the potential for burns to the skin, or damage to internal organs from breathing in gases given off by some chemicals. Note particularly where substances are transferred from one container to another, or where unidentified substances are stored and used!

4.13 Personal safety

We have to remember the hazards associated with working with the public and the threat of personal violence, especially when the person is working alone somewhere. In particular, consider staff who open up or lock up premises (often on their own), and those involved in transporting money or valuables on behalf of the firm. Include reference to staff who have to deal directly with animals too, as an element of possible injury.

4.14 Travel

Setting timescales for sales or delivery staff that are impossible to meet except by breaking the law and speed limits, and putting themselves and others at risk, represents a significant hazard that should be addressed.

Depending on your business type and structure, there may be many other types of hazard that you have identified, and certainly this is not intended to be an exhaustive list. It should, however, give you a good idea of what we mean by 'hazards', and Checklist 4.1 (p. 53) should help you to collate this information.

It is important to note that the legislation does not require you to write down the results of your assessment of hazards and risks unless you employ five people or more. Moreover, it says you must record the results of 'significant' findings, rather than every minor detail about every activity in your business that could possibly result in harm. It does NOT require you to carry out a risk assessment on the use of Tippex correction fluid!

Checklist 4.1 Safety hazards on site

No. (a)	Department or area on site plan where found (b)	Type of hazard found (c)	Type of injury or harm possible (d)

Chapter 5

Assessing the safety risks

There are so many different versions available to help you assess the health and safety risks in your workplace, that it is no wonder people become confused! The process is not intended to be a complicated, academic exercise that can only be carried out by a specialist, but a logical approach to identifying what the real risks are that hazards already spotted will result in harm or injury. There are some areas where technical or specialist help is needed, when assessing levels of noise or density of particles in the air for instance, but the main process of risk assessment can be carried out by you and/or a colleague in your own small business.

Having produced a comprehensive list of potential hazards in your workplace, you should now be able to take this a step further and consider the potential risks to people, based on

(a) who could be harmed;
(b) the severity of that harm;
(c) the likelihood that it will occur.

It is not necessary to give a numerical value to these evaluations, but it is sufficient to make a judgement based on criteria such as HIGH/ MEDIUM/LOW. So, using the list of Activities carried out in each area (Checklist 3.2) and the potential Hazards identified (Checklist 4.1) use the following headings to assess the risks (Checklist 5.1). It is easier to number the hazards as listed on Checklist 4.1, and enter the number in the first column of Checklist 5.1 (p. 58), rather than repeating them in detail although you will still have to look at them together.

5.1 Who could be harmed?

Which individual workers are in direct working contact with the hazard? It is useful to note whether these are exposed to the hazard most of the working shift, occasionally during the shift, on odd occasions, or perhaps once a year when maintenance is carried out. Include these details in column (b) in Checklist 5.1 (p. 58).

There may be other people too who could be harmed, such as customers or visitors, cleaning contractors or other businesses you share the premises with, so do not forget these people. While you might expect your own regular staff to be familiar with certain situations or processes in your workplace, don't forget that this is probably not the case with others.

You should also note any people who may potentially be more prone to the effects of some hazards, or less able to deal with situations or processes. These include young people under the age of 18 years old, whose lack of experience and expertise may increase the likelihood that injury or harm will occur, and older workers who may be extremely competent but who sometimes develop novel 'shortcuts' to processes over time! Statistics show that young men up to the age of 25 years old are most likely to experience accidents in the workplace, though not generally very severe ones. Older men, on the other hand, are less likely to have an accident but when they do it is likely to result in a major injury. Another group who are potentially at greater risk are nursing or pregnant women, who should be identified in your risk assessments.

5.2 Severity of harm

Don't forget that this is about the 'potential' for harm or injury associated with each of the hazards. It is a very subjective activity as people will differ so much in how they define the severity of harm, but you should be able to make a fair and reasonable judgement based on your own experience in the business. It will probably help to refer to any accident or sickness records you already have to remind you of the sort of injury that occurs in your firm, plus any manufacturers' guidance or information notes and the HDS we talked about previously. You should think about the type of injury possible so that you can then decide the level of severity.

The following headings can be used, or you can use others of your own to consider the severity rating against each of the hazards listed in column (a) and enter rating in column (c).

(i) 'Low' or 'Slightly Harmful'
such as minor cuts and bruises or superficial injuries that require first aid treatment.

(ii) 'Medium' or 'Harmful'
such as serious sprains or minor fractures, burns or concussion that result in lost time or hospital visits.

(iii) 'High' or 'Extremely Harmful'
such as major injuries, fractures, amputations and of course death.

5.3 Likelihood that harm or injury will occur

Again, there are many ways you can define this element, and by now you will already have a much clearer idea based on the work carried out above. So, based on the range of activities and hazards identified, plus your evaluation of what harm or injury is likely to occur and who is most likely to be affected, consider what the risk is that something WILL occur, then add this to column (d) on Checklist 5.1.

Remember that there are already some forms of control in place to safeguard people, whether it be guarding on machinery or protective clothing for workers (see next chapter for more details), or indeed safe working procedures in place for certain activities. The more people exposed to a hazard, and the longer the exposure time, the more likelihood that some harm or injury will occur. Another likelihood criteria could be:

High – if the activity occurs regularly, and it is already seen as a problem by workers or others.
Medium – something that happens perhaps once a month rather than daily or weekly.
Low – very irregular contact, perhaps for very short amounts of time in any given period, or a highly controlled activity with many safeguards already in place.

It may be a pretty depressing list by this time, seeming that every aspect of your work is potentially very hazardous! However, the point is that there are many ways to eliminate, reduce or control these risks some of which may be already in place, and some which will require very little additional action. You also cannot deal with everything at the same time, and some risks will be more significant than others or will need quite urgent attention.

5.4 Priorities for taking further action

As the law requires you to record the results of significant findings of risk assessments, there needs to be a decision or rating system in place to help you decide what is 'significant' or not. One way is to look at your results against the two criteria of (b) the severity of potential harm and (c) the likelihood that it will occur. Very simply this means plotting them on two axes (see Risk Table 1), with results ranging from a 'highly unlikely/low harm' point to the other extreme of 'very likely/extremely harmful'. This should result in a range from

1 = intolerable/urgent action required
to
5 = trivial or acceptable level of residual risk

While you may decide that little or no action is required to reduce the risks further at the 'highly unlikely/low harm' point (sometimes referred to as the 'Trivial Risk' point), this does not mean that you should just ignore your findings and expect people to accept the inherent risk without providing any controls at all. On the other hand, if you have several areas appearing in the 'very likely/extremely harmful' end of the spectrum, then clearly some urgent action would seem to be needed.

Risk Table 1 Assessing the risks

	Slightly harmful or Low harm	*Harmful*	*Extremely harmful*
Low likelihood or Highly unlikely	**Trivial risk** 5	4	3
Medium likelihood or Likely	4	3	2
High likelihood or Very likely	3	2	**Urgent action required – intolerable risk** 1

Add the number values to column (d) on Checklist 5.1 once you have decided which box in the table is most relevant to each hazard. Review these values after considering controls in place and whether they are sufficient to shift them down a rating where relevant.

Checklist 5.1 Assessing the risks

Hazard No:	Who could be harmed? (b)	Severity of the harm (c)	Likelihood that it will occur (d)

Chapter 6

Controls in place to reduce risks

We have already mentioned 'controls' that exist to safeguard workers or other people, and to reduce risk of injury or harm. While it is generally easy to see physical guards and controls that exist, especially in production or manufacturing areas, there are many other forms of control that exist and which you will already have in place. The assessment of risk you have carried out so far goes a long way towards demonstrating that you are fully aware of what happens in your company, how work progresses through the system, what the main activities are in each area and who does what. In addition, you have gone some way towards assessing the potential risks to people of injury or harm, and deciding which elements need further consideration.

Before you can decide exactly what this plan of action should involve, it is vital that you look carefully at existing controls overall, to identify just where any gaps are, and where new control features may need to be introduced. You may need to refer to other documents and guides you have to decide on the detail, but we shall cover the main points next. So, what do we include under the heading of 'Controls'?

6.1 Possible types of controls already in place

PPE – personal protective equipment

This will include things such as:

- protective clothing such as overalls; surgical type gloves; thermal wear and gloves; hats; hairnets;
- hard hats; toe-protection footwear;
- specialist wear such as rubber-soled footwear where there is a danger of electric shock;
- goggles; hard lens spectacles;
- chainmail aprons or gloves; reinforced wear (such as that containing Kevlar) for forestry work;
- harnesses for working at height or in confined spaces.

T & S – Training and supervision

Various things can come under this heading, including:

- training and qualifications in specific areas, such as manual handling; fork-lift truck driving; welding; use of hazardous substances;
- in-house training programmes at induction or on use of particular machines;
- direct supervision where hazardous activities take place;
- supervision to ensure PPE is worn where deemed necessary.

P – Procedures

- Regular checks to ensure good housekeeping standards are maintained.
- Specified safe procedures for carrying out tasks.

Figure 6.1 Good housekeeping is essential, especially keeping aisles free of obstacles. Note here the cloth hanging out at the bottom of the locker, and the container that overhangs the gangway lines.

- Formal 'Safe System of Work' procedures which rely on authorized personnel taking responsibility for ensuring systems are being followed correctly, particularly hazardous activities, for example, when machines are turned off for maintenance, or roof work is being carried out.
- Regular programme of testing and checking machinery or equipment, and recording results – essential for pressure equipment, lifting gear.
- Regular visual checks and testing of portable electrical equipment.
- Regular checks before the use of ladders, scaffolding, etc.

G – Guards and physical controls

There are many forms of physical controls, and it will depend on the type of industry you are in, but the following may be in place already:

- guard rails and covers for when machinery or equipment is in use;
- fail-safe systems to cut off power to machines in an emergency;
- stop buttons or mats;
- adequate alarm systems to warn users or passers by that some hazard exists, for example when vehicles are reversing;
- local or general exhaust and ventilation systems which are adequate and appropriate for the conditions;
- the use of PPE by individuals in certain areas on the site, or when carrying out specific tasks.

RA – Restricted access

It may be that access to certain areas, machines or substances are restricted to control the hazard, such as:

- security code locks on access doors to restricted areas;
- only trained personnel allowed to carry out tasks, for example driving a fork-lift truck;
- designated hard-hat or hearing defender areas.

These codes can be used to list the controls already in place, using column (b) on Checklist 6.1. It is also worth noting where there is a history of accidents or injuries, or indeed near-misses, even if they are quite minor in nature (in column (c)) before then deciding if they are sufficient to protect people adequately.

You now have to identify what further controls or actions are needed in order to ensure people are protected adequately if your review suggests they are not already. While the types of controls identified above are valuable in themselves, and are listed in the order that people are most likely to recognize them in their own workplace, they are actually in reverse order from the steps that the law says should be taken. You do, therefore, need to consider each of the following options in the stated order when deciding what actions are needed to reduce the risks to workers or others.

Figure 6.2 Identify situations on site where access is restricted to those wearing appropriate PPE, such as hard hats and boots in construction areas, and ensure that safety signs are displayed to remind workers and safety procedures followed at all times.

6.2 Required order of actions to be taken to establish effective and sufficient controls

(i) Elimination
(ii) Substitution
(iii) Restricting access
(iv) Physical guards or controls
(v) Procedures
(vi) Training and supervision
(vii) Personal protective equipment (PPE)

Elimination

Clearly the best way to protect someone from potential injury or harm is to remove the cause, that is the hazard itself. So, for example, if the use of a particular machine presents a particular hazard, then scrapping it altogether may be the most sensible option you can take. There may be hazards associated with using certain types of materials or fibres, or you may have identified hazards related to certain processes. Perhaps this is the time to consider alternative ways to carry out these processes, or to review investment or purchasing plans. In any event, as this is the most effective way to safeguard people, then it should be the option that is considered in the first instance.

Substitution

Having looked at the hazards and risks identified on Checklists 4.1 and 5.1, in the context of how your business operates, elimination of the source of the harm may just not be feasible. However, it could well be a more sensible option to consider substituting materials, machines or equipment with less-hazardous versions. For example, safer tools can be used to open cartons or boxes than short-blade hand-held knives, or materials may be available in different formats or strengths to safeguard the user. Of course, don't forget that you will need to re-assess the hazards and risks when a substitution is made, to ensure you have not introduced some new hazard.

As consumers and manufacturers become more aware of environmental issues related to use and disposal of materials, substitution for more eco-friendly materials or substances could well give commercial as well as safety benefits to your business. Contact your suppliers for further information on what or how you can substitute products, and include their responses in your evidence box.

Restricting access to the hazard

If you cannot remove the hazard, another option may be to remove people themselves from it. There are, of course, several ways to do this but again it will depend on what happens in your particular business. Is it possible to enclose or screen-off a machine or process that presents significant risks to the user or passer-by? In the case of potential personal injury from contact with the public in sensitive situations, is it possible to provide a physical barrier between them?

On the other hand, it may just be necessary to allocate specific responsibility to one person or 'key holder' so that you can control who has access at any given time. To some extent, this fits closely with the sections on procedures and physical guards, so the important thing is to identify what controls are in place, and what alternative forms of control might be needed. This is NOT about using a very precise definition, but about protecting staff and other people.

Physical guards and controls

Some equipment and machinery comes from the manufacturer complete with in-built physical controls or guards. Unfortunately, we all know how inventive workers can be when trying to by-pass these controls! As the business owner or senior manager, you are responsible for making sure adequate guards or controls are

- in place;
- in working order;
- appropriate for what they are guarding against;
- used correctly at all times.

Figure 6.3 Where necessary, fit specially made covers over cables trailing behind desks or across gangways and pedestrian traffic routes. Keeping computer cables organized is vital to reduce safety and fire hazards.

Another problem you may have is with old or inefficient machinery, and it is certainly more difficult to install new guarding systems to such machines. However, it is not impossible, and the development of such technology is rapidly changing. If you have identified that new guards, controls, or barriers need to be fitted, then you must take action on this. Again, contact manufacturers and suppliers for further information as the first stage of your actions (unless you have in-house expertise).

Procedures

Once you have considered all the options above, it is important to ensure the working procedures are themselves appropriate and safe for those following them. You will already have started to look at these earlier, and hopefully have identified where any shortfalls exist that put your workers at greater risk of harm or injury.

These need to be considered alongside points about individual workers, and any special steps that need to be taken to safeguard them, such as young workers or people with some disability that restricts certain activities. Other crucial elements of this section are shift or working patterns, rest periods, and the amount of repetitive tasks that make up their job. Tiredness is often a cause of accidents or near-miss incidents, and you should already have identified the potential areas in the workplace where this might arise. Do not forget schedules set for drivers on behalf of the business, especially as more and more towns have restrictive access times for commercial vehicles.

Training and supervision

This is not a substitute for having appropriate controls in place, but is a vital element in reducing risks and the likelihood that accidents will occur. It can often be a problem if you have a rapid turnover of staff, or you regularly use temporary or contract workers. It is easy to mistakenly believe that some aspect of the job is 'common sense', as such sense often only comes with experience or having witnessed the negative impact of taking certain actions.

You must ensure that people are adequately trained AND supervised so that they

- are aware of the hazards associated with the tasks;
- know the correct procedures to follow;
- understand why they must follow them in this way;
- can recognize early enough when things are not going well.

In addition, remember it is not enough to tell them to read a manual to find out how to work some equipment or machinery, you have to be confident that they DO know. Add notes to your Checklist where you think further training might be needed, and where supervision may need to be increased even if only for a short period.

Personal protective equipment

This is the last line of defence to protect workers, and should only be considered when other forms of protection or control are not feasible in the circumstances. Many firms do see this as the easiest way to control the hazards, and sometimes the cheapest, but it should only be used to offer very specific protection to an individual who cannot be fully protected by other means. Clearly, there are some circumstances where it is appropriate, where breathing apparatus is used for instance as other controls over the atmosphere are not possible. In addition, metal link aprons and gloves are used in the butchery trade as the use of knives cannot easily be eliminated or substituted!

There are many forms of PPE available, and manufacturers are happy to provide help and guidance. The main points to consider are:

- what is the specific hazard being protected against;
- whether the suggested PPE is appropriate for this hazard;
- that people know and understand how to use PPE correctly;
- it is maintained properly and replaced at suitable intervals;
- it is kept clean;
- it fits correctly, especially if more than one form of PPE is used at a time (for example, goggles and breathing mask);
- it is not a substitute for other forms of control.

6.3 Summary

This section is intended to help you move a step further than just spotting hazards and identifying potential risks to workers and others. While both those steps are a valuable starting point, you have to then take some action to make sure that if they cannot be eliminated, residual risks are controlled adequately. Note the controls in place and where they may be insufficient on Checklist 6.1 (see p. 67), using the numbered hazards from Checklist 1 in column (a) as you did when identifying risks.

You may have identified many areas of your business where the risks are significant, controls are in place but may not be sufficient or appropriate, and further action is now needed by you to deal with them. Until now, we have really concentrated just on safety and security issues which tend to be more easily spotted, and which people in the business are generally more aware of on a day-to-day basis. Clearly, you cannot deal with everything at the same time, and some actions may be quick/simple/relatively cheap, while others may need more substantial consideration or financial outlay.

Before you decide the order of actions to take, we need to carry out the same process to identify hazards and risks associated with both Health and Fire, so that you can first see where any combined action is more appropriate. Chapters 7–9 will cover Hazards, Risks, and Controls in relation to health issues. Chapters 10–12 will cover Hazards, Risks, and Controls in relation to fire issues. The same pattern of Checklists and questions to consider will be followed, so now that you are familiar with this process, it should be much quicker to complete these sections.

Checklist 6.1 Controls in place

Control codes: **E** = Elimination **S** = Substitution **RA** = Restricted access
G = Physical guards **P** = Procedures **T & S** = Training & supervision
PPE = Personal protective equipment

Hazard No: (a)	Existing control measures (b)	Details of any history of accidents (c)	Any gaps identified? (d)	Further control measures needed (e)

Part C

Controlling Health Risks

Chapter 7

Health hazards

In the previous three chapters we identified safety and security hazards in your business, using your own site plans to act as a prompt or reminder while you looked at every stage of the process. The approach you used when completing Checklists 4.1, 5.1 and 6.1 will be used again now in exactly the same way with Checklists 7.1, 8.1, and 9.1. Of course, you should be able to work through this section quite quickly as it follows the same pattern as Part B.

As with safety hazards, there are many different ways you can define 'health hazards', but from a practical point of view the following list of headings should cover the most common types of health hazard you and your workers or customers are likely to meet. Unfortunately, it is often much more difficult to see how far the way you and your staff normally work might threaten or damage an individual's health. This is particularly so when some situations result in recognizable health damage only after a very long latency period – just think how long it is after exposure to asbestos that actual lung damage appears.

However, it is clear that there is much greater awareness now of how damaging to health some work situations are, and as an employer you have considerable responsibility to ensure your staff are as safe and healthy as they can be. We have also included some environmental hazards and risks in this section, covering the most common elements you are likely to have to consider.

As with Checklist 4.1, use the site plans to work through the different areas of the business and to identify potential health hazards in each area, completing Checklist 7.1 at the end of this chapter with the findings.

7.1 Manual handling/back pain

You will already have started to look at this aspect of work in your business in relation to safety hazards, so should be familiar with where significant handling tasks take place. This does not just relate to lifting situations, although obviously these can potentially be very harmful for individuals, but also to pushing and pulling items, materials, or perhaps animals or people. The crucial thing is to look for twisting movements, or

Figure 7.1 This specially designed lifting aid is simple and efficient to use, especially when lifting heavy or awkwardly shaped loads. There are many devices available that will help to avoid injury and harm, but also reduce the likelihood of damage to goods and materials.

situations that require more than one handling movement at the same time, such as lifting and pushing a storage box onto a high shelf.

As mentioned earlier, the constant repetition of sometimes quite small movements can be very damaging over time, whether in a production setting or when using a computer or checkout system that relies on speed for its effectiveness. So many working days are lost through back pain that it is clearly something that all businesses will have to look at. However, there are lots of 'myths' associated with back pain especially in relation to how soon someone should return to work after a back injury. Some of these points will come out again in the following chapters on risks and controls, so for now you just need to identify the aspect of your work that could potentially result in some damage to the muscle or bone structures of people, and what that damage might be.

7.2 Noise levels

Hearing loss, even at fairly low levels, is very debilitating for those concerned and often takes place over a very long period of time before individuals notice it. People do get used to quite significant levels of background noise, but you certainly need to check levels where people regularly have to shout to each other as the norm in their workplace!

Damage can be caused either by long-term exposure to high levels of noise, but sometimes it can be just as harmful to be exposed to regular short-term blasts of noise over a working day. Constant repetition of certain types of noise can lead to distress in workers, or inability to concentrate, and sometimes the actual physical surroundings can make sound levels worse than they need to be – for example, when outer cabinet walls vibrate with the sound rather than absorbing and cushioning it. Evidence is starting to show that Callcentres, or situations where staff are required to use telephone headsets for long periods, can be very damaging to hearing capacity of individuals, so make a note of where this might be a problem.

Damage is likely to be short-term or long-term hearing loss, possibly leading to premature loss and deafness, or conditions such as tinnitus (a constant ringing noise inside the ear). Noise pollution is a broader problem that you may need to address, especially if some of the processes are particularly loud outside the building, and indeed outside site boundaries. Do they take place outside normal business hours 0900–1700? During night shifts or at weekends? Have they increased as the scale of operations or type of machinery has changed over time?

7.3 Lighting levels

This relates to both close and distance work, and particularly the amount of natural light that is available for people to work from. Look for areas where lighting causes shadows or glare on computer screens or other equipment, and especially in areas such as stairs, lobbies or entrance halls, and storage areas. The type of light does of course make a difference to how comfortable it is for people to carry out tasks, so note areas where low wattage, fluorescent or perhaps daylight bulbs are used.

Poor or inappropriate lighting can lead to headaches, eye strain, and pains in back or neck muscles. There can often be symptoms of depression or fatigue too.

7.4 Temperature levels

Mainly this relates to extremes of heat or cold, and there can be a dramatic change in working temperatures according to the season, so bear this in mind when identifying potential hazard areas. Ventilation may be an issue, and consider whether people are working inside, outside, or in areas such as chill houses or freezers. If temperature levels

Figure 7.2 This dental repair workshop is well planned, with adequate storage at convenient heights and all main tools easily to hand. Extra adjustable spot lighting is anchored firmly so that it does not tip.

are not correct for the type of work being carried out, there may be lower levels of dexterity and concentration, sometimes difficulty in breathing normally, and sometimes loss of consciousness.

7.5 Air quality

Check that existing extraction or ventilation systems are adequate and working, whether local or general systems, or areas where they do not exist at present but may be required. The level of dust particles in the air may be fairly easy to see, but the tiny particles that are virtually invisible to the eye may be even more hazardous and just appear as clouds of 'mist'. Also remember to include where people work in confined spaces, and areas where exhaust fumes from vehicles present a potential hazard.

Poor air quality may lead to fairly minor coughing, nausea or eye irritation, but could be much more hazardous in its effect on breathing and respiratory actions as well as causing drowsiness or symptoms similar to being intoxicated. You should also identify where extraction and ventilation system exhaust fumes are emitted; the volume of extraction; and how hazardous these exhaust fumes might be to surrounding areas and the environment.

Figure 7.3 Exhaust and ventilation systems, even simple ones like this, need to be checked, cleaned and maintained to keep them working efficiently. If they have been in place for some time, also check that they are still in the right location for the best effect.

These last three categories of physical working conditions of light, temperature and air can certainly have a negative impact on morale and motivation, which may result in dissatisfaction with the job and therefore loss of concentration and commitment, so where possible you need to get it right.

7.6 Use of computers and visual display units (VDUs)

In this section you should note where the VDUs or screens are positioned in relation to keyboards and seating, and ensure that everything can be adjusted to suit the individual user. As before, note whether wrist rests are used and people have had training in correct procedures, lighting is appropriate to avoid screen glare, and people have proper rest breaks. Types of harm or damage likely are similar to those listed for lighting levels above.

7.7 Micro-organisms and airborne contaminants

If your business is in food preparation, you will probably already be registered with the Local Authority and be following the detailed requirements of HACCP (see Industry specific case studies in Part F). The following are potential hazards that many businesses may meet. Include food preparation areas, even if only used by staff, looking at things like surfaces, temperatures in fridges, food reheating facilities, washing facilities. There may potentially be airborne contagious diseases from clients, customers, other staff, or in areas where you are working with animals. Specific contaminants such as those associated with blood products should already be known by you, and identified on the Checklist, and do not forget disposal of products in skips or other refuse collection facilities.

Disposal of such contaminants is also an environmental issue, and appropriate systems must be in place to ensure proper isolation and labelling of such waste, and that appropriate collection procedures are used. If unsure of the level of contaminants, contact the local Environmental Health Office for guidance.

7.8 Radiation

Again, this is a very specific hazard that you should already be aware of if it is an issue in your industry sector, but you may have to consider whether any low-dosage radiation is potentially a hazard to female workers who are (or could be) pregnant or nursing mothers. There may also be potential concerns about exposure at even low levels and the effects on fertility levels. Environmental damage by emissions into the atmosphere must be monitored, as must the disposal of possibly contaminated waste products.

7.9 Use of chemical and other substances

Includes the storage/use/disposal of cleaning agents and other chemicals, especially solvents/adhesives/inks/dyes/mineral oil substances. In hairdressing there are many bleach-based or otherwise hazardous substances to identify, and a wide range of chemicals used in agriculture.

The potential harm can be eczema or dermatitis type skin conditions, which are often extremely debilitating and usually long-term or recurrent. As with some airborne chemicals, they can also act as 'sensitizers' for severe effects in the future when an individual comes into contact with the substance, and of course there may be some substances that are carcinogenic (that is, cancer causing).

Environmentally friendly methods of use, storage and disposal must be in place. Note in particular where substances are stored; evidence of leakage into ground or water courses; how waste products are disposed of; how contaminated materials are disposed of; where empty containers are stored for disposal (still with traces of chemicals inside).

7.10 Use of materials and fibres

Identify any of the production or storage areas where dusts and fibres can escape into the air, such as agricultural preparations and fertilizers, wood dusts from sawing/sanding, cement and similar products in construction, flour dusts in bakery areas, and of course asbestos. Many of these

Figure 7.4 Chemical substances used in carpentry workshop. Apart from checking that chemicals are clearly labelled and correctly stored in the right containers, periodically check whether less toxic substances can be substituted to reduce the potential for harm.

dusts are potentially cancer-forming and most can lead to respiratory, digestive or skin diseases. Note also storage, leakage, or disposal of such materials.

7.11 Smoking

Identify areas where smoking is specifically restricted or allowed, and ventilation systems in place. You should have a policy for smokers and non-smokers in your business (see later Part E), as you need to consider the impact of smoke on non-smokers, and consider the atmospheric conditions in work and rest areas. This is a difficult issue, as there may be a question about how far you can protect workers in, for example, the leisure industry where you may not be able to restrict customer smoking.

As well as the serious damage that can occur to lung and throat tissues, there may be irritation and discomfort experienced by non-smokers, so air quality is an important issue.

7.12 Work organization

Include shift patterns worked, schedules and targets set, and rest periods. There is a lot of evidence now that the way work is organized can have a significant impact on how 'stress' is handled. It is extremely difficult to be precise about causes of stress, especially as people react differently to the same situations. However, you do have a duty to protect workers' health, and the feelings of not being able to cope effectively with a situation generally relate to being set unrealistic targets, with insufficient equipment or materials to complete the job properly; tight deadline; insufficient training or chance to learn the skills properly; and too much individual responsibility. It is worth noting that the owner of a small business may well see stress in a different way from workers, but on the other hand workers are not paid to take on the same responsibilities as the owner!

If people are not allowed to take their rest breaks during or between shifts, this can clearly lead to tiredness and lack of concentration. In addition, bullying or violence at work either internally or in contact with customers can be a significant factor in the levels of well-being experienced by staff.

7.13 Summary

There may be other potentially hazardous situations that you can identify in your business, and as we said previously, using the site plans should ensure that you have considered all the relevant areas of the firm. Note all the results on Checklist 7.1 (p. 79), numbering each hazard as you go along so that you can refer to it more easily on the next two Checklists to assess the risks and identify the controls.

Checklist 7.1 Health hazards on site

No. (a)	Department or area on site plan where found (b)	Type of hazard found (c)	Type of injury or harm possible (d)

Chapter 8

Assessing the health risks

The process you will follow here is similar to that in Chapter 5, 'Assessing the safety risks', where you considered:

(a) who could be harmed;
(b) the severity of that harm;
(c) the likelihood that it will occur.

Use the Risk Table on p. 57 for assessing health risks, again considering where is the most appropriate place for each hazard using the given criteria. However, in the case of potential health risks, there may be some issues that need expert input to identify the exact level of hazard present, and the potential severity of the harm it could cause. Use Checklist 8.1 at the end of this chapter in the same way as Checklist 5.1, adding the details in the relevant columns.

8.1 Who could be harmed?

Which individual workers will be exposed to the hazard identified, and which other people may be exposed such as customers or passers by? Note also the times when the hazard is most likely to be present, for instance during certain weather conditions for temperature and light levels; or it may only be certain stages in the process when the potential hazard exists, such as during cleaning or at disposal.

With hazardous substances, it is often contract or part-time staff most likely to be exposed to the hazard, and this could well be outside normal business hours for you. In addition, your workers may be using some preparations off-site when they visit customers at their own premises, for example. These situations may still require you to have some controls in place to protect people, so you must include them in your risk assessment.

It is even more important with health issues to take into account the characteristics of individual workers, such as their age, sex, susceptibility to certain conditions or ailments, and previous health history. Again, young workers under the age of 18 years should be specified in your

Checklist, as should pregnant or nursing women. In some circumstances, you may also have to consider exposure to any staff whose fertility levels could be affected.

8.2 Severity of harm

As noted previously, this is about the potential for harm if not controls are in place to protect people. When considering health hazards, you are more likely to need some professional input to help you decide how potentially damaging the hazard can be. This is particularly the case for noise levels, air quality and airborne contaminants, and radiation levels.

Testing and recording noise levels is a fairly simple and quick process that is available widely, although testing individuals' levels of current hearing loss will also need to take place. However, once you know these details you can identify anyone who already has some damage that will be made worse by further exposure, and what levels of protection are needed in the future (see Chapter 9).

Testing the extent and nature of air contaminants is again a specialist task, but should also be widely available to businesses nationally. The starting point should be discussions with workers in the area to identify the current levels of discomfort or harm experienced by them, and how severe they think this is, as clearly they are the best placed to know. Also consult sickness absence records where they exist, and accident records, to see if any patterns emerge. Enter your severity rating against the hazards as listed on Checklist 8.1.

8.3 Likelihood that harm will occur

Using the same criteria as stated in Chapter 5, consider the hazards identified and the people likely to be affected, alongside the judgement about how serious the damage to their health or well-being could be, to decide what is the likelihood that harm WILL occur if sufficient controls are not in place to protect them. Remember, the less time people are directly exposed to the risk, generally the less likely they are to be affected by it.

Use Risk Table 1 again to consider where you would place the hazards according to the risk factors identified above, and enter the findings on Checklist 8.1 in column (d). If preferred, you can use the number value in each box to enter details on the Checklist, but remember we shall be reviewing these values later in relation to the controls you have in place.

8.4 Priorities for taking further action

As with safety hazards and risks, we shall consider priorities for action when all three Parts B, C and D of the Guide are completed, to ensure the actions complement rather than conflict with each other.

Checklist 8.1 Assessing the health risks

Hazard No. (a)	Who could be harmed (b)?	Severity of the harm (c)	Likelihood that it will occur (d)

Chapter 9

Health controls in place to reduce risks

Having completed the Safety Controls Checklist and identified the different types of methods or activities that exist to help control hazards, and reduce the likelihood that people will be harmed by them, you should be familiar with the process by now. However, as we have already seen, health hazards are not always so easily recognized and many of the controls in place may be taken for granted, and indeed assumed to be more effective than they might actually be in practice.

Use Checklist 9.1 to help you organize the details about hazards identified, existing controls, and any further control features you may need to introduce. We shall use some of the same headings as in Chapter 6 to identify the possible types of control already in place, so you can use the same abbreviations on the Checklist.

9.1 Possible controls in place to reduce risks

PPE – Personal protective equipment

In addition to those listed for safety controls, these might include:

- protective clothing, hats and gloves;
- disposable single-use protective wear;
- ear plugs and ear defenders;
- spectacles for use with VDUs or other equipment;
- masks that cover nose and mouth, with range of different quality filter pads to fit;
- more complex breathing apparatus that ensures a flow of breathable air and covers more of the face or head;

- hands-free telephone equipment, but also note the comments about Call Centre staff;
- personal monitors and alarms;
- wrist or body supports for lifting activities.

T & S – Training and supervision

Including measures such as
- training and qualifications to handle or work with specified processes;
- training in storage, handling, disposal of very hazardous substances;
- training in how to handle potentially violent situations, including self-defence;
- adequate training for lone workers who may be particularly vulnerable;
- training in proper use and care of PPE;
- direct supervision while particularly hazardous activities taking place;
- supervision to ensure PPE is worn in designated areas.

P – Procedures

Day-to-day procedures in place such as

- proper use of Hazard Data Sheets from suppliers;
- regular programme of maintenance and oiling working parts of machinery, to reduce noise;
- organization of work for habitual (regular) VDU users, to ensure adequate rest periods and breaks from using the screen;
- good housekeeping and cleaning regimes maintained;
- designated Smoking and No Smoking areas;
- appropriate decontamination procedures and adequate washing and toilet facilities for people;
- adequate rest periods and breaks to reduce stress or discomfort levels, and suitable rest areas;
- provision of eye tests, and spectacles where necessary, for habitual VDU users;
- first aid and other training to ensure proper help and support in emergency situations, such as breathing difficulties or spillages of contaminants.

Monitoring and surveillance procedures could also include

- regular checks on dust, noise, temperature and contaminant levels in relevant areas;
- regular hearing tests to monitor any increase in hearing loss over time, perhaps for specified staff;
- records of monitoring health of staff exposed to potential health hazards such as air contaminants, radiation, or micro-organisms.

Figure 9.1 Good housekeeping in vehicle repair premises is particularly important where hazardous substances are concerned, avoiding containers left open; different hazardous substances kept together in corners instead of in properly labelled storage units; naked flame ignition sources next to flammable liquids or aerosol sprays.

G – Guards and physical controls

There may be fewer of these in relation to health hazards, but the main ones are likely to include

- muffling or cushioning on machinery casings to reduce vibration and noise levels;
- meters and other recording systems that are regularly checked and maintained, perhaps with in-built alarms to warn of unsafe levels;
- closed containers for moving or storing potentially hazardous substances;
- provision of adequate cleaning and washing facilities with correct cleaning preparations in use.

RA – Restricted access

These are likely to be similar to those listed for safety controls, including

- only trained personnel allowed to carry out specified tasks;
- security code locks on access doors to restricted areas;
- sufficient notices available to warn people they may be approaching restricted or hazardous areas;
- designated hearing defender or protective clothing areas (that are supervised and enforced!);
- restricted access to personnel already identified as more susceptible to the hazards identified.

Use these codes as before to identify the controls already in place, and again do check your own internal records to see whether they have actually been adequate in the past. As with the safety controls identified earlier, you now have to identify where the gaps are in your current provision, and consider what further actions need to be taken.

9.2 Required order of actions to be taken to establish effective and sufficient controls

(i) Elimination
(ii) Substitution
(iii) Restricting access
(iv) Physical guards or controls
(v) Procedures
(vi) Training and supervision
(vii) Personal protective equipment (PPE)

Elimination

The first option is to eliminate the hazard where possible. You may find that in relation to health hazards, there are some situations that can be eliminated quite easily, particularly those related to work organization. For instance, many potential back injuries can be avoided by changing the position or height of work stations, shelving and packing stations. Placing VDUs to suit the individual user can also eliminate some of the potential harm, as can altering the work patterns where possible. The issue of habitual users taking a short break away from looking directly at a computer screen has been misrepresented to some extent, in that it does not require the person to go out and have a cup of coffee for 5 minutes in every half hour! The break is intended to give a rest from looking directly at the screen, so the user can regularly spend a few minutes involved in other work at the same station – see also the section on Procedures.

Potential air contamination from dusts can sometimes be eliminated just by changing the structure of the materials used, for instance being supplied with a granular or pellet form of the material rather than a fine powder. Health hazards associated with smoking can be eliminated by introducing a complete no-smoking policy throughout the firm. This may be appropriate in some businesses, but can be a sensitive issue and needs to be discussed fully with all the people concerned. Allowing smoking in designated areas can be a more effective option (also note the comments in the next section on fire risks).

Substitution

This could well be a realistic option for your business if you have identified hazards related to the use of chemicals, other substances, materials and fibres, as many of these can be substituted with less hazardous versions. You will need to contact suppliers for more information if this is an option for you, and again include copies of product information in your evidence file. As we noted earlier, substitution for more 'eco friendly' materials could be what your customers or clients are looking for so can represent significant business benefits for you.

However, it is not always a viable option, as technological developments may not be at the stage where substitute products perform the same tasks as effectively, or give such a good quality result. This has been the case in the plastics industry with water-based inks, and with some cleaning products. If you cannot eliminate the hazard altogether, then this is the next option that must be considered before relying on methods listed later. It can, of course, be a realistic option in relation to the type of lighting or ventilation system currently in place, so does need to be considered against all of the hazards identified by you.

Restricting access to the hazard

The most obvious health hazards that can be controlled in this way include radiation and airborne micro-organisms, and access to some of the chemical substances. In some situations, larger noisy machines can be isolated within separate rooms or compartments, and access restricted to specified personnel or those wearing appropriate ear protection. It may be that supply of the equipment relies on this provision. Locked cages and cabinets for storing some materials may be a simple but effective option, and could fit with the controls needed for fire risks later in the guide (in Part D).

In other cases such as welding booths, there is no reason for people other than those working there to have access at all, although clearly this needs to be monitored and supervised. In practice, there are often good basic controls in place but they are allowed to lapse over time, so this is a good time to be reviewing them. For instance, the use of unmarked containers may be used to transfer substances from larger containers, and the control relies on people in the organization knowing what the practice is. You may quite rightly decide that this is insufficient!

Physical guards and controls

In a similar way to restricting direct access to the health hazard, physical means may keep the individual removed sufficiently from the hazard to considerably reduce the likelihood that they will be harmed. In this context, there are many physical guards that can protect people from noise, radiation, and other contaminants. These often rely on appropriate containers being identified and used correctly, which also relies on people being trained and supervised properly.

There are antiglare screens available for VDUs, and temporary screens that can be fixed over existing units. There is some doubt that radiation levels from VDUs pose any significant threat to health at all, and you should contact the HSE for guidance before buying products that are supposed to offer protection (see Sources of advice and guidance in Chapter 17).

Procedures

We have already noted a wide range of procedures that would help to safeguard individuals from possible harm, and you may need to introduce further systems to ensure adequate protection is provided. This may just be in relation to specific groups of people you have identified as particularly vulnerable, or as a result of introducing some of the additional measures discussed above.

What you do need to do is review procedures to ensure that they are still

- relevant and appropriate;
- workable;
- being followed correctly.

It may be necessary now, having identified particular health hazards such as use of some fertilizers, to set up a formal health surveillance or monitoring system for specified individuals; check whether this is required for your industry. This does not automatically mean that a complicated, time consuming system needs to be set up, but you do have to show you are taking adequate and effective measures to protect workers and others.

Training and supervision

This is a vital step to ensure all the previous actions taken to reduce the potential harm to the lowest feasible level are effectively established and maintained. It does not necessarily have to involve sending people off-site for training, as what you need is relevant training. It does not necessarily mean that it will cost lots of money either, as many Local Authorities provide training on general health and safety issues, and many manufacturers provide training on the use of their products. However, it may involve you finding out what is available locally, and establishing a training plan for staff.

If there are only five or six people working in your firm, then this should be fairly straightforward to organize. There may also be other small businesses locally who need to take similar actions, so working together may be a realistic option to keep costs down.

In any event, if new processes or procedures are introduced, people will need to know what they are and will need to be adequately supervised until they are familiar with them. Your evaluation of the hazards, risks and controls should show you exactly where this action is needed.

Personal protective equipment

As you know by now, this is the last step in the protection hierarchy after reducing the risks as far as possible by other means. However, there will still be situations where this is a vital part of the protection regime and indeed the action needed may be to up-grade existing PPE.

Manufacturers can give you advice and guidance on the most appropriate PPE for your circumstances, and the last three Checklists provide exact details of the situations where it is needed. We have already noted many points about the choice and use of PPE, but it is worth remembering that for it to be effective it must be:

- appropriate for type of hazard;
- able to provide sufficient levels of protection for the worker;
- correctly fitted according to the individual's physical build;
- compatible with any other form of PPE they have to wear at the same time;
- clean and stored/maintained in proper manner;
- replaced when necessary to maintain the level of protection.

9.3 Summary

As with the steps taken in Part B – identifying hazards, risks and control measures for safety – you have now looked closely at the potential health hazards and risks facing workers or other people, and have considered the measures that are possible to control these hazards and reduce the risks.

Checklists 7.1, 8.1 and 9.1 should now be complete and form part of your evidence file. However, as we noted at the end of the last section, before you decide on the priorities for action we need to include one more element of risks facing people in the business, that is the fire risk. Although the process will be similar to that you have done already, some of the headings will be different. Part D, Chapters 10, 11 and 12 will look more closely at fire hazards, risks and controls.

Checklist 9.1 Controls in place

Control codes: **E** = Elimination **S** = Substitution **RA** = Restricted access
G = Physical guards **P** = Procedures **T & S** = Training & supervision
PPE = Personal protective equipment

Hazard No: (a)	Existing control measures (b)	Details of any history of accidents (c)	Any gaps identified? (d)	Further control measures needed (e)

Part D

Controlling Fire Risks

Chapter 10

Fire hazards

While most people are familiar with health and safety hazards, it is sadly the case that unless you have witnessed the speed and destructive power of a fire at first hand, you are unlikely to be fully aware of what we mean by fire hazards and risks. The local fire authority is always willing to offer advice about your own particular circumstances, and although you might need to call on some professional assistance at some stage, the majority of your fire risk assessment can be carried out by you and your workers.

Recent changes to the law mean that even if you only employ one person, you now need to carry out a fire risk assessment in the same way that you assess the health and safety risks. As you have probably realized by now, although it sounds and looks more complicated than it actually is when written in a guide like this, the process is a fairly basic one. You also have a valuable starting point with your site plans, as much of the fire risk assessment can be carried out by reference to these plans.

10.1 Plan D – Site plan to show fire hazards and preventive measures in place

In Part D, we shall produce a further site plan which identifies

- internal and external areas where fire hazards may exist;
- fire fighting equipment and alarm systems that are in place;
- escape routes for people in the event that a fire has started;
- areas where action is required to reduce risks to people.

If you already have a Fire Certificate or other form of approval from the fire authority, you may still need to do this risk assessment, although it will probably be a fairly simple step to identify if any further controls are needed. Another important point to remember in the case of fire risk assessments is that areas of shared ownership or use of buildings need to be identified, and other people notified of your findings.

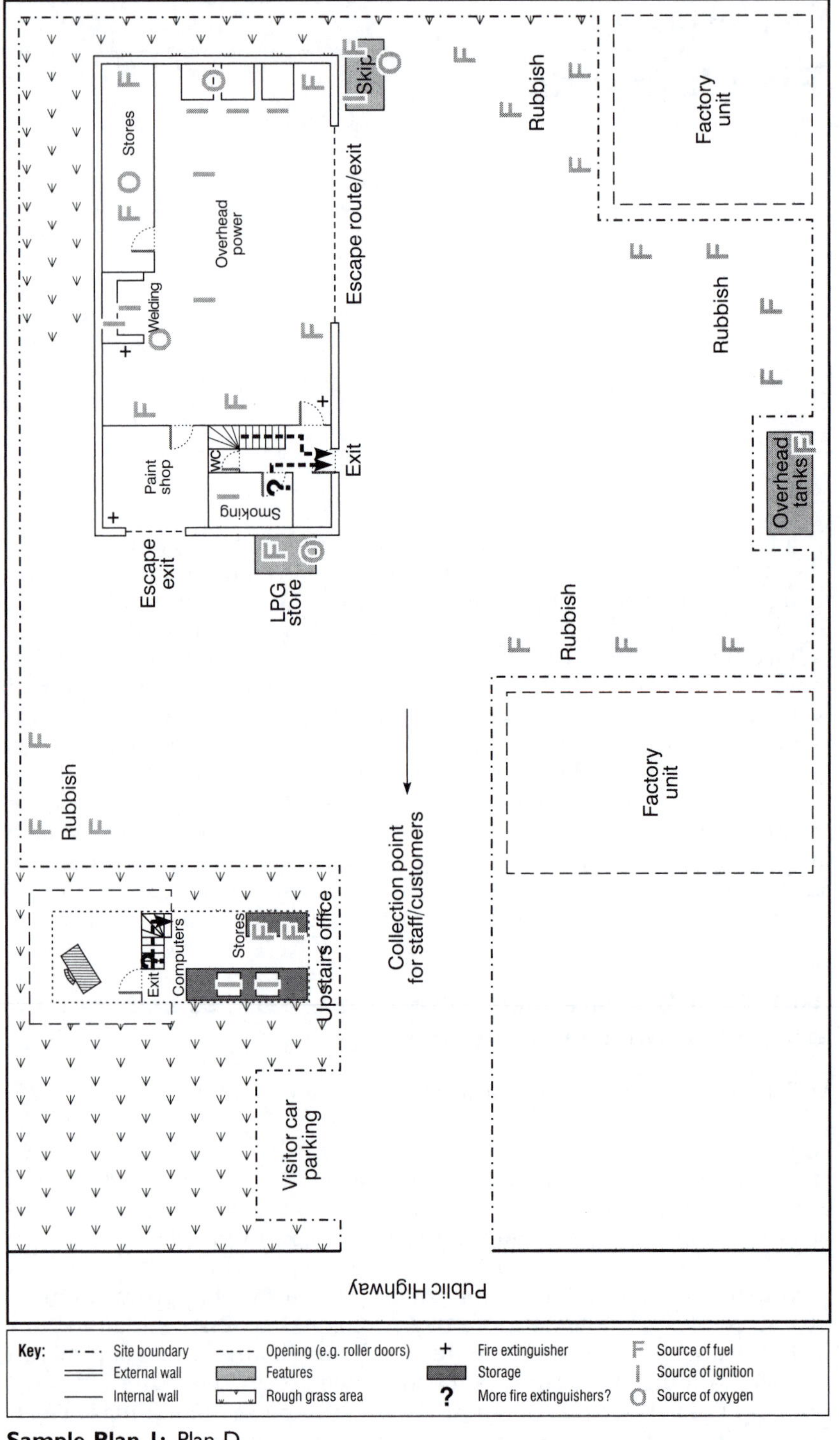

Sample Plan 1: Plan D

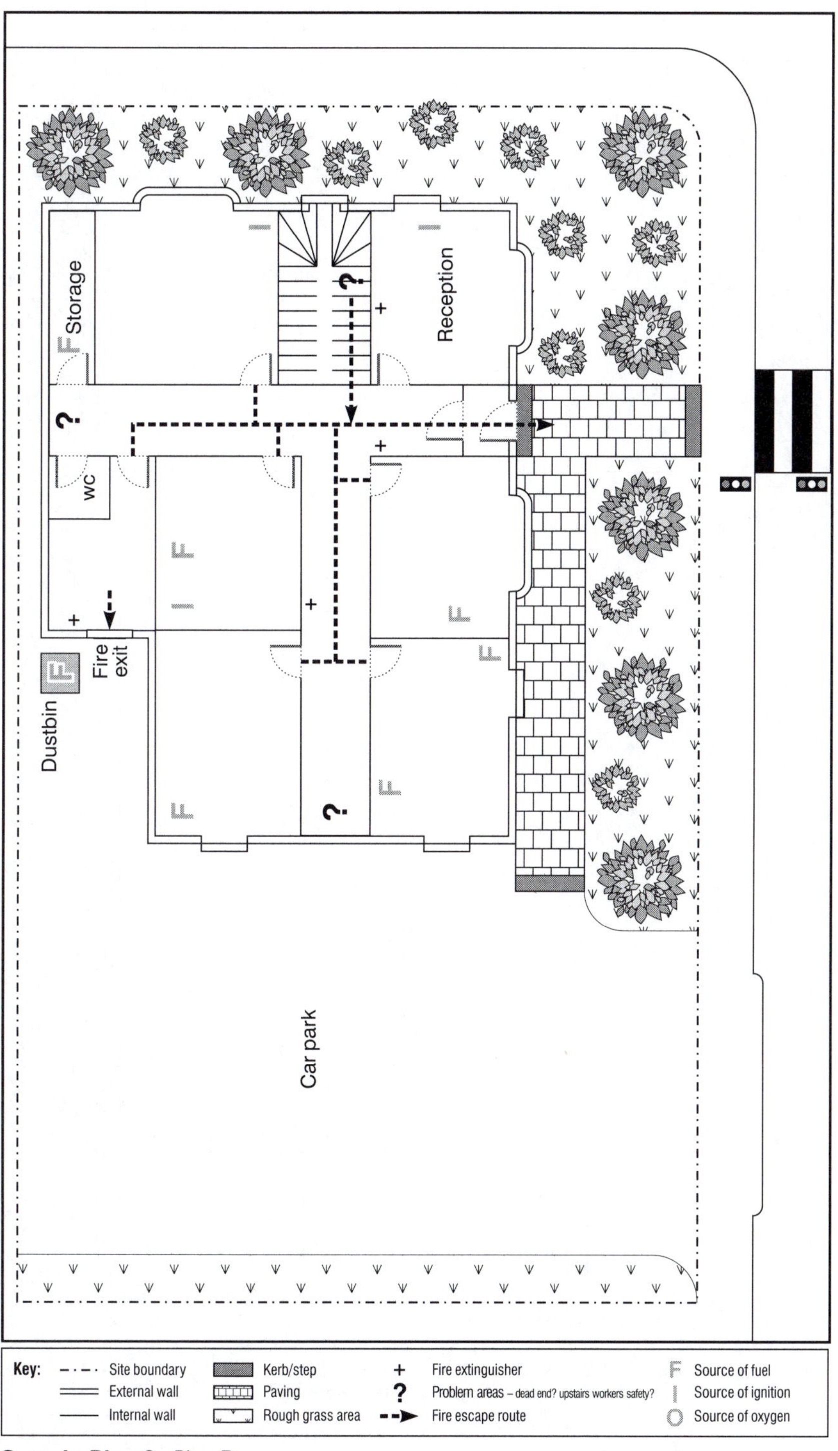

Sample Plan 2: Plan D

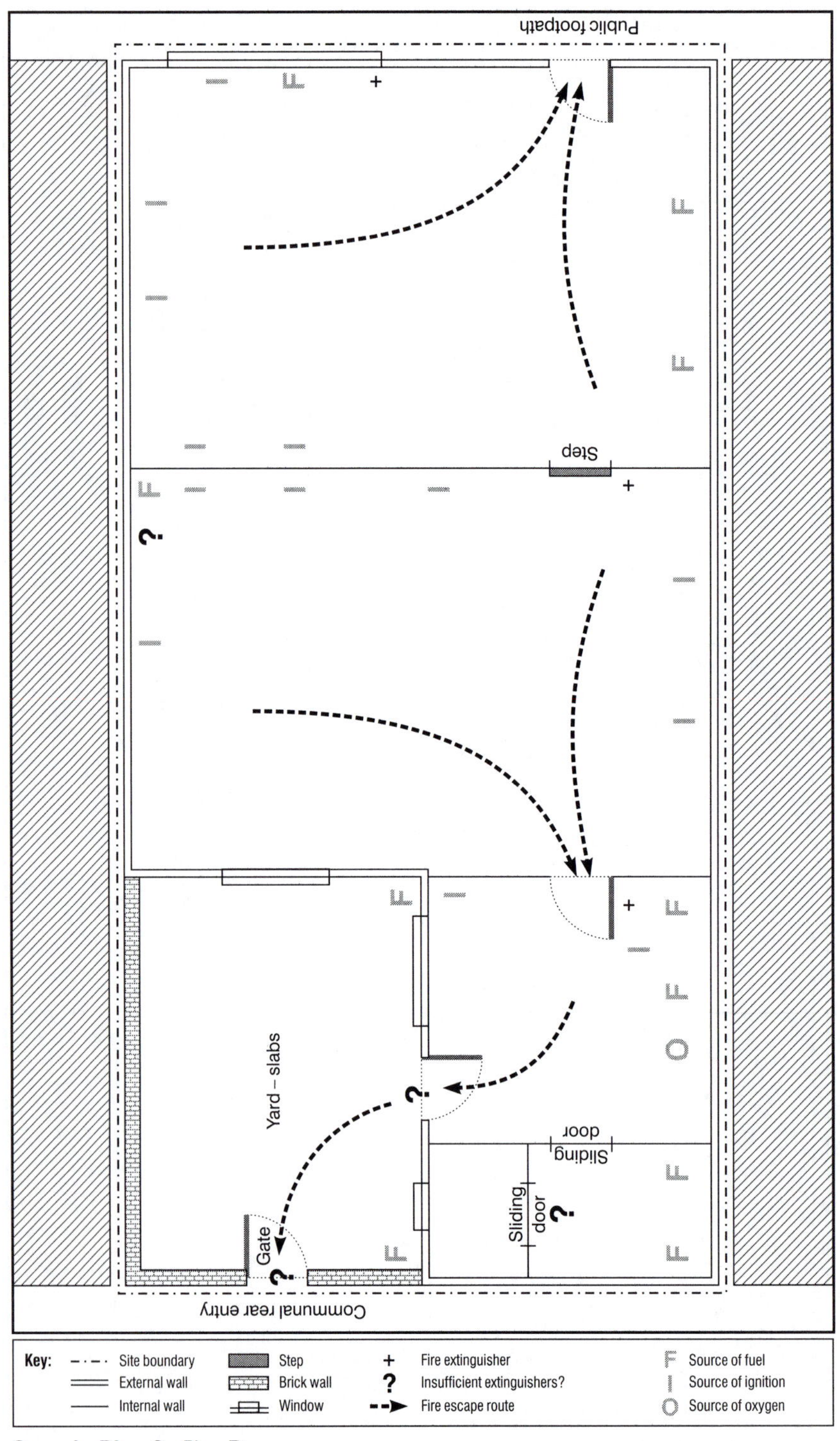

Sample Plan 3: Plan D

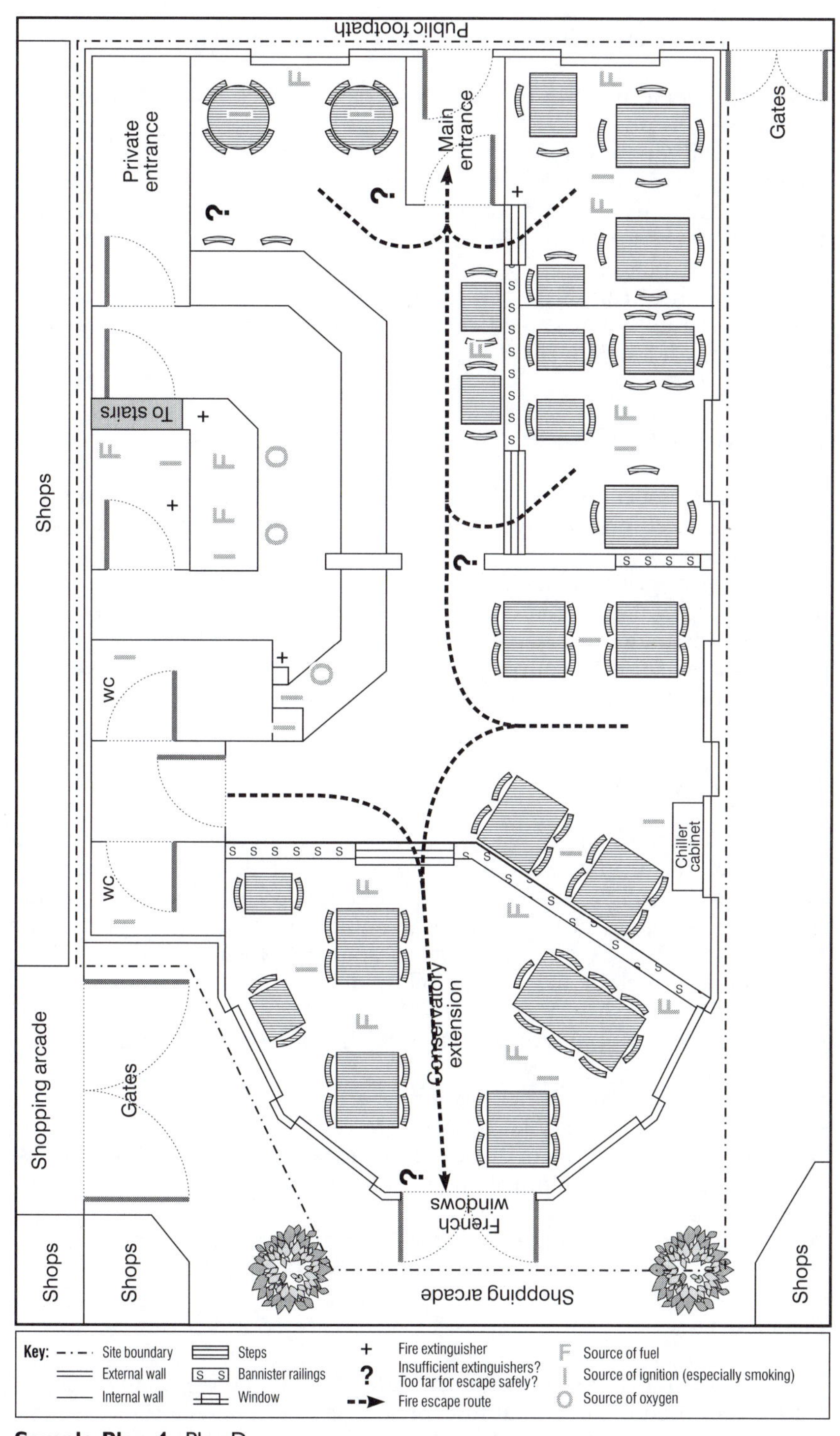

Sample Plan 4: Plan D

Points worth remembering before we start to identify fire hazards are:

- smoke from a fire rises to the ceiling, gets trapped there, then spreads wherever it can;
- smoke will quickly spread to other parts of the building through any holes or gaps it finds;
- smoke can be extremely toxic depending on what materials are burning;
- heat from the fire intensifies and can cause materials or substances to ignite or explode;
- flames or heat can 'leap' across to other buildings or structures;
- fire and smoke will spread rapidly in open-plan areas, roof cavities, corridors and stair wells.

10.2 The fire triangle

The hazards associated with fire are the three elements it needs in order to burn – sometimes referred to as the **fire triangle** – which are:

- a source of **ignition**, like a spark, naked flame or hot surface;
- **fuel** to keep it going such as flammable gases and liquids, or other flammable materials;
- **oxygen**, which is of course present in the air but may also come from chemical substances or in pressurized containers.

Sources of ignition

Using the outline Plan D, which shows outside areas as well as internal features, work around all the different areas of the building/site and identify with an 'I' where you find potential sources of heat or ignition. It is worth looking out for places where you know fires have previously taken place or been avoided due to someone's quick actions. These sources might include any of the following:

- areas where people store, smoke or dispose of cigarettes, matches, cigarette lighters;
- heaters using gas, electricity or oil as a fuel;
- naked flames, pilot lights, cookers, etc.;
- machinery for welding or grinding, or processes which produce sparks;
- smaller pieces of equipment such as lights or office equipment where surfaces get hot;
- faulty electrical equipment or areas of high static electricity.

Fuel sources

This will include anything that burns fairly easily, and especially where it is used or stored in large quantities. These should not just be sources that are likely to help start a fire, but also the fuel that could feed a fire once

Figure 10.1 Powerful lights such as these are a potential fire hazard, especially if covered by diffuser materials, placed near flammable materials like curtains, not secured safely or with trailing cables likely to trip someone.

started and keep it going for longer. Again, using Plan D to make sure you cover all the areas where your business operates, identify with an 'F' the potential sources of fuel. Do not forget to include waste disposal areas outside the buildings too. These sources may include

- paper, cardboard and packaging materials;
- wood;
- furniture and shelving or other fittings;
- furnishings and fabrics or similar fibres;
- foam, polystyrene, polyurethane or similar products, whether they are part of your production process or part of the building structure itself;
- chemicals and solvent based products, especially petrol or spirit based products;
- use of paint, varnish, adhesive products;
- gases such as LPG and acetylene (generally found in cylinders).

In addition, look at the processes carried out in different areas, and note those where dusts and fibres are released into the air. This could include wood, paper or fabric fibres but may also occur from other materials such as flour, cereals and animal feedstuffs. Depending on how densely they

Figure 10.2 Space is often at a premium in small retail outlets, so it is even more important to consider where and how products are stacked to reduce the likelihood of rapid spread of fire once it starts.

float in the air, they could certainly help to spread the fire quickly from its original source without proper control systems in place. The way your building is constructed or laid out internally may itself present some fuel hazard once a fire has started, so if you are uncertain about this then contact the local fire authority for advice.

Sources of oxygen

Obviously, we need oxygen in the air we breath, so unless your workplace is in a hostile environment (such as underwater or underground) this source will be present in sufficient quantities to fuel a fire. In some buildings, the ventilation system will ensure this oxygen is moving freely around the building and will even add more where necessary. You do not have to add this to the plan unless there is an obvious air flow route around the room/building which could increase the likelihood that a fire will keep burning, or if there are known means for shutting down the flow. You could, however, note with an 'O' external windows and doors that open and can provide additional oxygen to the room or building.

Some of the processes you carry out might involve using pressurised containers, air or oxygen cylinders that could potentially add to the spread of a fire. Of course, there may also be some chemicals you use that act as oxidizing agents (you should already know which ones these are as the details should be included on the labels/manufacturers instructions for use and storage).

10.3 Summary

Alongside Plan D you can use the Checklist 10.1 (p. 102) to note the hazards you have identified around the site, and which processes take place where. These areas of activity, and the main people involved in them, have already been identified earlier, so should not present any great problems. It may be that while the potential hazards individuals meet in certain parts of the building are not great, this is not the case with potential fire hazards.

In this case, we also have to consider areas which might be particularly vulnerable to arson attack, and to the careless disposal of cigarettes. Skips and waste containers present further problems if materials that are individually safe are mixed with another element of the fire triangle. Few people realize the potential for a fire to start inside a bin of waste rags soaked in inks or solvents seemingly without a source of ignition being added. In the next chapter, we shall consider the level of risk you are exposed to in your firm based on these initial findings.

Checklist 10.1 Fire hazards on site

Note: Use this as a summary list or prompt when identifying these details on Plan D.

No: (a)	Area where hazard identified (b)	Sources of ignition? (c)	Sources of fuel? (d)	Sources of oxygen? (e)	How many people? (f)	Controls identified? (g)

Chapter 11

Assessing fire risks

In Chapter 3 you used Plans B and C with Checklist 3.2 to identify what activities take place in each area; in Chapter 4 you used checklist 4.1 to identify the potential safety hazards that exist in each area, and who is likely to be exposed to them. In the same way, you have used Plans B and C with Checklists 3.2 and 7.1 in Chapter 7 to consider health hazards and who could be exposed to them, although the individuals identified may have been different in each case.

While we shall be taking the same approach in this part to consider the fire hazards that exist, the question of who might be at risk is perhaps less clear. Fire risks are much more closely tied to location rather than specific activities, although of course these have an impact too. While the potential for harm or injury to occur to an occasional visitor to a particular area of the site is relatively low, these occasional visitors may in fact be at greater risk from fire than someone who is familiar with the site.

This means that your assessment of risk may need to be more carefully considered in this part than in the previous two. In addition, you may well need to give greater consideration to the potential risks to customers, contractors, and even unauthorized visitors who could be on site when a fire occurs.

It will probably be easier to consider both Plans C and D alongside each other, as well as Checklist 10.1 and the one included in this chapter, Checklist 11.1, to make sure you get a full picture of the people at risk. This should include other businesses that share the premises with you, and especially areas of communal or joint use, such as entrance halls, stair wells, rubbish disposal points.

11.1 Who is at risk

Using the plans and checklists as prompts or reminders, look at each area or section of the business including outside the buildings, and note the following:

- which individuals regularly work in that area, and note if there are shift changeover times;
- identify relief staff or people who cover for breaks;

- consider who has to use the area for access, or as a route between sections and departments;
- include reference to customers or clients that may be there at any particular time, even if only for short periods;
- where sub-contractors, temporary or agency staff are used, especially if they are likely to bring additional fire hazard materials with them;
- where 'work experience' or young people are likely to be working;
- passers by who might be at risk if a fire started.

In addition, note where there are likely to be individuals who could experience difficulties with escape if a fire starts, such as

- elderly or frail people;
- wheelchair users or those with restricted mobility;
- pregnant women, parents with small children, or indeed unaccompanied children.

You should also make a note of areas that are rarely used, are largely unoccupied, or where a fire could go unnoticed before the alarm was raised. Just as importantly, you must also consider where the most vulnerable areas are for arson attacks, especially near perimeter fences and where flammable or scrap materials are stored.

11.2 Severity of harm

In relation to fire, the severity of harm that could occur is principally the same in most situations, and cannot readily be related to whether the person is carrying out one activity rather than another. If a fire starts and people cannot escape in time, the result could be burns, asphyxiation or being overcome by smoke and toxic fumes, or major injury while trying to escape from above ground level.

The severity of harm could be related to other things, such as whether a lot of people or just one or two individuals could be harmed; whether there is a danger of explosion; or the impact a fire could have on other businesses in the vicinity. It is vital that you know if adjoining businesses use highly flammable or volatile substances or procedures, and that they have also assessed the risks and have sufficient controls of their own in place.

The following may be a more useful way to consider the potential severity of a fire if one started:

- a small-scale fire that could be tackled safely in the early stages by a competent person; very localized;
- a localized fire to start with, but could spread very quickly;
- a localized fire to start with, but one that could quickly produce toxic smoke and fumes;
- a fire that would almost instantly spread over a wide area, perhaps via a spread of chemicals or other highly combustible materials or dusts;
- significant potential for explosion.

11.3 Likelihood that a fire will occur

Closely linked with the potential severity of harm is the likelihood that it will occur, and what steps you have taken to reduce this likelihood. The main considerations are

- whether a fire is likely to start;
- what controls are in place to reduce the chances of it starting;
- how will people know it has started;
- and how can they escape.

While the next chapter will look in more detail at the different ways to control and reduce the risks, it is useful to give some sort of risk rating to the potentially hazardous situations you have identified, as we did for health and for safety. While the question of whether it is Unlikely/ Likely/Very Likely that a fire will occur is similar to the criteria used in previous risk assessments, the criteria for the severity of harm in Section 11.2 above will be a more practical approach.

Figure 11.1 In vehicle repair workshops, as in other types of business, there are often many factors present that increase the likelihood of a fire starting, with many potential ignition, fuel and oxygen sources close to each other.

11.4 Priorities for taking further action

The chart will now change to look something like Risk Table 2, with more columns to show the potential severity of the fire, and additional criteria of number of people likely to be affected. The rating becomes more or less significant according to the number of people involved.

Risk Table 2 Assessing the fire risks

	Slow burning localized	*Rapid spread localized*	*Toxic smoke localized*	*Rapid spread widely*	*Explosion*
Few people affected	5	5	5	4	3
Highly unlikely	5	5	4	3	2
Likely	5	4	3	2	2
Very likely	4	3	2	2	1
Many people affected	3	2	2	1	1

This is just one way you might give some rating value, and on this basis you would need to consider ratings of 1–2 as extremely high risk, 3–4 as medium risk, 5 as low risk (though still needing attention of course), and these can be amended to take into account vulnerable groups already identified. As we have already said, it is a very subjective way to consider the risks, but does give you a starting point. You might just use the table to consider the risks against these factors, but without giving them a number value.

Checklist 11.1 Assessing the fire risks

Include sector headings for groups of people who may be at risk, such as staff; customers; visitors; temporary workers; passers by

Area on site	Main activities carried out	Who is at risk? (sector and number)	Specific difficulties likely?	Severity of fire	Likelihood a fire will start

Chapter 12

Fire controls in place to reduce risks

Following the same procedure as previously, use the Checklist 12.1 to note the existing controls in place to reduce the risks to people if a fire starts, and later to identify further actions needed where controls are insufficient.

12.1 Existing controls

These will vary according to how big your business premises are, type of industry you are in, and lots of other variables, so could easily be a fairly short list. The main things likely to be in place include:

- **A – Alarms**: fire, heat or smoke alarms – this could range from a handbell in very small workplace to raise the alarm, to electrically operated alarm systems that warn people on site and alert the fire service at the same time. The main consideration is whether people in the workplace can be warned quickly enough to escape safely if a fire starts. Your Plan D and the last two checklists should also have alerted you to areas where a fire could start and remain unnoticed for some time, such as in storerooms or areas that are generally unoccupied. In this case, you might also consider whether an automatic alarm system of some kind is needed, or whether it is still sufficient for a manually operated alarm to be in place.
- **S – Signs**: signs and notices – you should have notices displayed to tell people what to do if a fire starts and they hear the alarm, even if these are very simple notices. Signs that tell people where fire exits are, and the quickest route to them, should obviously be clear and visible and in a language that is appropriate to the people working on site. Remember that if there is a lot of smoke, people easily become disoriented and frightened to move through it to safety, so it is vital that people know where to go in this situation.
- **E – Exit routes**: exit routes – note where fire doors and fire exits are located. In particular, make sure doors are not locked, are easily and quickly opened in an emergency, and that they do actually lead people to safety! These exit routes and doors must be kept clear of rubbish,

goods or other obstructions at all times, so check that this is actually the case. They must also be lit well enough for people to see clearly (you might need some form of lighting that does not rely on the main electricity supply). If your assessment identifies people with mobility difficulties could well be on site, these exit routes should be able to accommodate wheelchair users too.

- **FFE – Fire fighting equipment**: fire fighting equipment – remember, there should be enough equipment to enable people to fight a fire SAFELY during its early stages. This could include appropriate fire extinguishers, fire blankets, or hose reels close to where potential sources of fire exist, as identified on Plan D. There may also be sprinkler systems in your premises, perhaps just in specific areas of risk. In any event, make sure that all equipment is regularly checked and maintained, and get professional advice if you are not sure whether you have the right equipment available.
- **P – Procedures**: procedures and 'fire drills' – everyone who works on your premises should be aware of the procedures in place to escape the building in the event of a fire starting. They should be given information, trained in how to use equipment, allocated responsibility for making sure procedures are followed, and responsible for keeping exit routes and doorways clear of obstructions. They should know how to raise the alarm, and how to help others evacuate the premises if necessary

12.2 Further controls that may need to be introduced

While the measures above are a valuable part of the way you control the risks from fire, there may be other measures you can take that give better protection to the people you are responsible for. Some of them may already be in place, but some may need to be introduced or developed further. The three main elements of control are:

(i) **RD** – reducing the likelihood that a fire will occur.
(ii) **EP** – preparing for an emergency.
(iii) **AA** – taking appropriate action if a fire has started.

Reducing the likelihood that a fire will occur

The question to consider is 'how likely is it that a fire will start in a given area as the three parts of the fire triangle (ignition source, fuel, oxygen) are, or could be, present together?' The ways to reduce this likelihood include:

- Controlling sources of ignition, identified by the 'I' on Plan D. Are heat-producing machines or equipment maintained properly, and ducts or flues kept clean? Electrical equipment should also be checked, circuit breakers used where appropriate, and socket points not overloaded. You may also need to replace heaters that have a naked flame with convector heaters or central heating systems – also check with your insurers if you use portable gas heaters on the premises, as your insurance cover may be affected by this.

Any smoking policy or procedures for 'hot work' such as welding or flame cutting should be enforced properly, especially the use of matches or lighters. Checking procedures at the end of the working day, or in areas of low occupation should also be in place, and can be a significant control factor.

- Reducing potential fuel sources, identified by the 'F' on Plan D. There are many ways to control this element, often by fairly simple changes to the way materials are stored or handled. While you might replace some substances or materials by less hazardous ones, as we discussed in the previous chapter, changing the way they are used in the process can also minimize the risks. For instance, by only keeping the smallest workable volume at the point of production, rather than very large amounts, and transferring materials in a safe manner; by using fire-resisting cabinets to store highly flammable liquids and substances, and keeping them separate from other flammable substances where necessary. Good housekeeping and proper disposal of waste products could also significantly reduce the potential risk, especially the opportunities for arson attacks.

Figure 12.1 This is an example from a carpentry workshop where a mix of highly flammable substances that can potentially lead to explosion and rapid spread of fire are stored next to a good supply of wooden planks that will keep the fire burning for longer!

- Reducing potential sources of oxygen, identified by the 'O' on Plan D. Simple actions like keeping doors and windows closed where appropriate, and checking that oxygen or similar cylinders are stored safely and with proper ventilation. You will already be aware of substances you use that act as oxidizing agents, so make sure they are stored away from heat sources and flammable substances.

Preparing for an emergency

You can take precautions like those listed above to try and reduce the chances that a fire will start, but unfortunately fires do occur for many reasons so you also need to make sure that you and your workers know what to do if it does. The procedures you have in place could already be sufficient for your premises, and you need to check that they include:

- Taking into account your identification of potential hazards, where they exist, and who may be affected, you have established a realistic and effective 'plan of action' that will safeguard people. You have already identified areas where there may be particularly vulnerable people, so the plan should take their specific needs into account. You have also looked very closely at the way the business operates, and where everything is located, so again your plan should take into account the physical aspects of warning people and escaping from the premises if a fire starts.
- In a very small business, it will be quite easy to make sure all the people involved know what to do if there is a fire, but you should check that this is the case especially where you have new, temporary or contract workers on site. It may seem unnecessary, but you should also test that the plan of action actually works, and that people do remember what they have been told. This is especially important where you have customers or vulnerable groups of people on site.
- It is also important to make sure that your emergency procedures are known by others who may share the work premises, and do not in fact conflict with procedures they have. Don't forget that you should also be informed about their plans, and especially any significant fire hazards they have identified that could pose a serious risk to your business.
- It is vital that the emergency plan includes reference to checking and maintaining equipment, alarms, and exit routes or doors. It should also refer to checking at close of shift, and security measures we looked at in earlier chapters to reduce the opportunities for arson attacks and for unauthorized people to be on site.
- As with all procedures, people must be trained adequately, and training up-dated as required, whether it is in safe systems of work, using fire fighting equipment, or first aid training. It should also note details about shutting down machinery in an emergency, checking that all personnel are safely outside the building, and contacting the emergency services.

Taking appropriate action if a fire has started

This element is largely covered in the section above where you have identified the controls you already have in place, as quite often this is the most obvious part of the procedure, and the one people are more aware of. It is primarily about warning people that there is a fire, ensuring they can escape safely, and preventing the spread of fire where possible.

- Make sure the alarm can be raised, and that people can see it and hear it, particularly if it is a noisy workplace or individuals have sight or hearing difficulties. If there are only a few people on the premises, perhaps just one main working area, it could well be sufficient to raise the alarm by shouting 'FIRE' or by using a handbell. The crucial thing is that the system is adequate.
- Ensure escape routes and exit doors are adequate, clearly signposted, and well lit. The exits must, of course, lead people to safety so check that this is the case whatever the time of day you are working. For example, if your fire exit leads to an outside courtyard that is open to the public during the day, but perimeter gates are locked for security purposes after dark, people would be trapped in the event of a fire.
- Simple actions that prevent the spread of fire include keeping the door closed where a fire has been sighted, turning off electricity supplies if necessary (but not if it then puts people in further danger), keeping fire doors closed, and making sure that materials or substances that could help to spread the fire are not stored too closely together. The building structure itself may be such that it helps the spread of fire, as may the furnishings or furniture, so these may need to be assessed separately.

12.3 Summary

Although this section has been a little different from the two previous ones, the principles are the same in that hazards have been identified, risks to people have been assessed, and control measures considered to reduce the risks as much as possible. This has been in relation to the size and type of business you have, so the plans and checklists you have completed for your business should provide a good base of evidence to demonstrate that you have assessed the risks and identified the actions needed to control them.

This process can be used to assess risks in relation to other aspects of the business, such as a more in-depth environmental risk analysis, or food hygiene risks. If this is relevant for your business, then use the same approach to carry out risk assessments of these aspects and add the results to your evidence file. In Part E, we shall look at how you manage the risks in the business, and all the general management procedures you have in place, as well as your Health and Safety Policy Statement.

Checklist 12.1 Fire controls in place

Note: **I** = Ignition sources **F** = Fuel sources **O** = Oxygen sources

Control codes: **A** = Alarms **S** = Signs and notices **E** = Exit routes **FFE** = Fire fighting equipment **P** = Procedures and drills **RL** = Reduce likelihood **EP** = Emergency planning **AA** = Appropriate action

Area where I, F, or O found	Existing controls	Details of history of fire or near-miss	Any gaps in control identified? If so where	Further controls needed

Part E

Effective Management of Fire, Health and Safety Risks in the Business

Chapter 13

Managing the risks

Up until now, this has been an auditing process to identify just where you are at the moment in relation to health, safety and fire risks to you and your business. You now have a file that contains information about the context within which your business operates, which includes:

- the type and structure of the business;
- the products you make or service you provide to customers;
- how the premises are laid out and activities organized in the business;
- who and where people are.

In addition, you have used this information, plans of the site and premises, and a range of checklists (also in the file) to identify how well you are controlling risks in the workplace by:

- identifying health, safety and fire hazards in each area of the business;
- identifying the people likely to be injured or harmed by these hazards, plus any individuals or groups of people that may be particularly vulnerable;
- assessing the potential risks to workers and others who could be on site at any particular time;
- reviewing existing controls that are in place to reduce the risks, and identifying further control measures that may be necessary.

While this is an excellent beginning, and demonstrates to others that you are taking these risks to health and safety seriously, we have not yet:

- decided any future targets or objectives for you and the business;
- considered their order of priority;
- produced a plan to take them forward;
- or identified ways to check whether you have met the targets or to measure the success of your efforts.

These elements of 'managing' the risks successfully are just as important as those of clearly identifying the context of the business and the risks people face within it. We have avoided specific reference to the law so far, although the activities you have carried out are, of course, based on what the law requires you to do. However, you are required by law to actually *manage* health, safety and fire risks to workers and others in your business, so must take the work done so far still further forward.

As you know by now, we have tried to keep the paperwork to a minimum, relying on visual plans, checklists, and existing literature you already have in the firm. The people we identified in the Introduction do, however, want to see *evidence* that you are actively managing health and safety effectively (relative to the size and type of firm you have), even though the law only requires you to record most of these things if you employ more than five people. I am sure you have also realized by now that despite this, it is clearly in your interests to have some record of actions taken and your commitment to the principles of good fire, health and safety management.

13.1 Priorities

We shall use Checklists 6.1, 9.1 and 12.1 together to get a better overall picture of where actions are needed to reduce risks further, which process or activity areas are involved, which people are affected, and any areas of overlap between the three risk factors that need to be addressed.

(i) First pick out all the factors you decided were *high risk/likelihood* for health, safety and fire on Risk Tables 1 and 2 plus your checklists.
(ii) Against each one in turn, check whether controls were assessed as adequate, and if so make a note to 'Review' later in the last column of the checklist. This review should take place when conditions change significantly, or in 12 months at the latest.
(iii) If the controls were not seen as adequate, then check this activity or process on the other two checklists to see whether more than one aspect needs attention.
(iv) Check that the actions already identified as necessary to control this risk do not conflict with each other on different checklists, and put *priority rating 1* in last column on each (we are not deciding actual order that actions will be taken yet, just giving them a rating for importance).
(v) Follow the same procedure to pick out *medium risk/likelihood* factors identified.
(vi) If existing controls were considered as adequate, make a note in the last column to 'Review' later.
(vii) If existing controls were assessed as inadequate, again check the activity or process on other checklists.
(viii) Put *priority rating 2* in last column.
(ix) Follow the same procedure to pick out *low risk/likelihood factors* identified.
(x) If existing controls adequate, make a note in the last column to 'Review' later.
(xi) If existing controls were not adequate, check against the other two checklists.

(xii) Put *priority rating* 3 in last column.

(xiii) Note that any factors considered low risk or trivial may not necessarily be assessed as perfectly controlled, so will still need to be reviewed in the future and might still need some action such as refresher training for operators, or new signs, etc.

13.2 Plan of action

You now have a priority listing of

- 1 = Urgent attention required, do as soon as possible.
- 2 = Keep a close watch on the situation, take action as quickly as possible.
- 3 = Keep a close watch on the situation, plan what action is to take place.

You can now set targets and produce a plan of action to show what you intend to do. Use Checklist 13.1 as a guide, and list all the actions needed starting with all the priority 1 factors.

There may, of course, be just two or three things that need to be done, or quite a long list of fairly urgent measures that are needed. You cannot do everything at the same time, and some may require significant investment of time or resources to put right. You might also need to get specialist professional advice on things like noise or air contamination levels before you can decide what the most appropriate actions will be. It is quite likely that there will be a list of steps that need to be taken before you can meet a particular target. For instance, if new storage facilities are needed, because of either a fire or safety hazard, you may need to:

- get professional advice from somewhere about exactly what it is you need;
- look at a range of suitable products available from different suppliers;
- decide which one will be right for you, in consultation with others in the workplace and suppliers;
- place an order;
- prepare the area where new storage will be installed;
- install the new facility;
- then reassess the situation to confirm that controls are now adequate.

By looking at all the actions required at the same time, you should be able to see where one supplier could perhaps provide several of the controls you need, rather than using several different suppliers. Having got to this stage, you may also be able to get help and advice from your local HSE, Local Authority, or Fire Inspector, or your own insurers, to check that the proposed actions are sufficient. A word of caution – if you receive conflicting advice from different inspectors, then ask them to visit your premises at the same time and agree between themselves what is acceptable BEFORE you commit yourself to any major investment.

Checklist 13.1 Managing the risks

Management actions taken:	Yes (tick)	In part	Complete by	Review date	Review by
(a) Established priorities: • noted 'Review' where controls adequate • identified high risk factors with priority rating 1 • identified medium risk factors with priority rating 2 • identified low risk factors with priority rating 3					
(b) Prepared Plan of Action, with steps needed and timescales set for completion, for • priority rating 1 • priority rating 2 • priority rating 3					
(c) Established appropriate records					
(d) Provided staff with relevant and sufficient information					
(e) Established appropriate consultation procedures with workers					
(f) Identified one or more 'competent person(s)' for Health & Safety					
(g) Arranged methods for keeping up to date with legislation changes					
(h) Prepared a Health and Safety Policy					
(i) Included other policy statements where appropriate					

You do have a duty to assess the risks and have adequate controls in place, so you do have to *do* something to follow up your assessments. As with any business targets, proper planning with timescales and clear measures to judge success are essential. Regular review of situations is also essential, of course, so you must also identify timescales for reviewing health, safety and fire risk assessments to make sure control measures are still appropriate and effective.

Further actions that form part of this section are:

- Assess potential hazards and risks if you change the product or process used, or if new equipment or technology is introduced.
- Make sure that new equipment or machinery is safe when you buy it, and suitable for what you intend to use it for. The CE mark is a useful guide, but you still need to check with manufacturers that it does not introduce other hazards such as excessive noise levels.
- Ensure that safe and healthy procedures are established when introducing new equipment, processes or materials, and that potential fire risks are identified.
- Make sure that people know and understand what the targets are in relation to health and safety, and how important their role is in making sure they are met.
- Make sure that working safely is accepted as relevant and 'the norm' in the business, with you/the owner of the firm setting an example (and sticking to it!).

13.3 Keeping records

Apart from the details you have collected so far in your evidence file, there are other records that you should keep. These include:

- Maintenance checks on machinery or equipment. In some cases, you are legally required to keep such checks, for example for lifting gear and hoists; regular checks on portable electrical equipment; and checks on fire fighting equipment. Check with HSE Infoline or other reference sources to see if any of these apply to your particular industry.
- Results of noise level and hearing tests, airborne contaminants, etc.
- Hazard Data Sheets that tell you about using and storing hazardous substances (which come from suppliers of such substances) need to be kept.
- Accident and first aid records must be kept, especially details of serious injury, diseases, or dangerous occurrences that have to be reported under RIDDOR (see below).
- Sickness absence records should give you additional information about which staff in which parts of the business are most at risk of injury or harm. They are also needed if sickness benefits become due to individuals.
- Details of accident investigations carried out internally should show what happened to whom, with what result, and where controls were insufficient to protect the person concerned. It is in your interests to

include near-miss incidents too if you can, as they often suggest where there is 'an accident waiting to happen' which then becomes a foreseeable event. The aim of accident investigation is to get to the root cause of the incident, to make sure it does not happen again. It can also identify potentially vulnerable individuals, areas where further controls need to be introduced, and situations where existing controls need to be reinforced. It is, therefore, an important element of showing that you are managing health and safety effectively.
- Records of individual workers' skills and expertise, plus training received and planned for the future.
- Names of qualified first aid staff and where they are generally located, details of those trained to use fire fighting equipment, and individuals responsible for acting as fire wardens or notifying emergency services, etc.
- Health monitoring and surveillance records for individual workers, where necessary (remember these are confidential and should be kept securely).

13.4 Informing and involving staff

This is a crucial part of the legislation in Europe and the UK, so should not be ignored or left to chance. Steps you need to take include:

- All workers, including temporary, part-time, contract staff, must be informed about potential hazards, the risks of injury or harm associated with them, and controls that are in place to safeguard individuals. As you have seen already in Chapters 4, 7 and 10, it is important to involve people working in the different areas on site when identifying hazards, as they will be more familiar with them on a day-to-day basis than someone from outside.
- If new hazards or risks to health are identified, you must tell people and make sure they know and understand the safeguards that are in place to protect them. It is not enough to just tell them to read a notice or leaflet.
- By law, you must consult with staff on issues of health and safety. This does NOT mean you have to have a formal Health and Safety Committee structure in place in a very small, close-knit firm where it is obviously inappropriate. However, you DO have to have some method for talking to workers, discussing health and safety issues or concerns, and agreeing future actions to ensure the safety of all. This could be by talking directly with all workers together as a group, or in a larger company by discussing issues or concerns through an elected workers' representative. You must remember to include off-site workers or those working outside normal business hours too.
- Relevant warning or information notices must be displayed where necessary, clearly and in a language or format that is easily understood by the people they are aimed at. You are obliged to display the HSE 'Health and Safety Law Poster' and details of public and employer liability insurance cover you hold.

13.5 Competent people

There is still no clear definition of what a 'competent person' is in the context of health and safety, but it is certain that someone given specific responsibility for dealing with some area of fire, safety or health management should:

- know and understand what they need to do;
- have the technical skills to be able to do it;
- have sufficient expertise in the subject area to be able to carry out the tasks to the required level;
- have sufficient resources and the authority as well as the responsibility to do it.

By now you have seen that many aspects of risk assessment do not need to be carried out by highly qualified health and safety professionals, as this guide should have given you enough help to carry out a large amount of the work yourself. In addition, it is often preferable to use existing internal staff where possible rather than buying in external professional help unnecessarily.

On the other hand, there are some parts of the process that you cannot do without the relevant expertise, such as eye tests, noise and hearing assessments, assessing levels of contamination, identifying potentially hazardous properties of substances you use, or individual health surveillance. In addition, people need to be trained in first aid and fire fighting techniques, and in some cases hold specific qualifications for using certain equipment or machinery. There are a variety of qualifications available in health and safety, but many are aimed at professionals rather than people taking on the responsibilities as part of other roles. In any event, people do need some sort of guidance and training if they are to take on such responsibilities.

The owner/Managing Director of the firm has ultimate responsibility for ensuring the safety, health and welfare of workers, and cannot escape this by making someone else the 'nominated person with responsibility for health and safety'. Every individual in the firm has some responsibility, whether as an employee, self-employed contractor, or as part of the management team. You have to make sure, therefore, that everyone knows this and is given relevant, appropriate training and support to be able to carry out such responsibilities. This includes proper induction training when they join the firm or come onto your site (this includes temporary workers), plus adequate training in the correct procedures for carrying out their job in a safe and healthy manner.

13.6 Keeping up to date

This is one of the biggest concerns for smaller businesses such as yours, especially as there is so much pressure now for the employer to take more practical and financial responsibility for the damage, injury or ill-health that results from work activities. It is complicated still further by the

Figure 13.1 There are many ways to keep up to date with changes in health, safety, environment and fire legislation ranging from suppliers' literature, newsletters and journals, specially produced guides and leaflets, to subscription based up-dating services.

increasing overlap of legislation between health, safety, environment, fire, employment protection and public health.

You have a duty to make sure you comply with health, safety and fire legislation relevant to your business, so therefore have to ensure that you know what the relevant legislation actually is. This could be a huge task, and could easily keep someone in your firm in full-time employment for the whole year! Clearly this is not a realistic option for most firms, so you will have to identify the sources of information that are most accessible and relevant to you.

Chapter 17 gives details of further reference sources you can use, and information services you can contact directly. As with this guide, it is not necessarily the detail of the legislation you need to know, but rather the underlying principles and purposes, and the actions you need to take to ensure compliance.

Chapter 14

Your policy

By law, you must have a policy on health and safety whatever size firm you have, but only need a written version of it if you employ five or more people. However, as we have seen already, there are likely to be other groups of people who want to see evidence that you have a policy in place, such as your insurance provider or a client, or when placing a tender for contract. Having got this far through the guide, you might as well record the main features of your policy in writing!

14.1 Drawing up a Policy Statement

In fact, you already have the details of your policy in the evidence file, so this section should be fairly brief and straightforward to complete. It starts with a general statement of your policy in relation to health and safety, and can be a list of bullet points like the following examples. In this case, the eight Ps from the Introduction to this guide are used as a prompt list to make sure all the elements are considered.

Example Policy Statement for company x

We are committed to:

- providing a safe and healthy work environment for people
- producing a product that does not jeopardize the safety and health of others, or the environment
- purchasing less hazardous raw materials where possible, that are healthier, safer and more environmentally friendly to use
- ensuring processes are carried out using equipment and machinery that is appropriate, as safe to use as possible, and properly maintained
- ensuring premises are maintained properly, good housekeeping standards are kept, and adequate facilities are provided for workers and others on site
- establishing procedures for work in all activity areas of the business that take into account the health and safety protection of workers

- making sure that procedures intended to safeguard people and the environment are followed correctly, and that people are adequately supervised
- ensuring suitable monitoring and recording systems are in place
- identifying hazards and assessing risks to workers and others who may be affected by activities of the firm
- providing adequate protection for people against the risks of damage to health, and harm or injury, resulting from work activities or fire
- involving workers directly in discussions about health and safety issues or concerns, to ensure their input and commitment to working together to tackle these issues
- providing sufficient resources, information and training to people to ensure they can carry out their duties and fulfil their responsibilities in a healthy and safe manner
- setting targets to reduce where possible accidents and ill health in the workplace
- regularly reviewing the situation to see whether targets have been met, existing controls are still adequate and in place, or new targets need to be set.

As you can see, this is an extremely comprehensive list with a lot of detail included. It would serve as an excellent reminder to you and your workforce about how far-reaching health and safety management is within the business, and how important it is to work together. However, it could just as easily be shortened with some of the points merged together, and some of them summarized – for example, you could just say 'providing sufficient resources, information and training to people' without saying why. However you set it out, it needs to be signed by senior people in the organization, and dated, with a review date included such as one year ahead.

Included with this statement of commitment you should also include references to named people in the firm who have specific responsibilities, details of how and where they can be contacted, and contact details for external sources of advice, support or services. These should include those responsible for carrying out risk assessments, maintaining machinery, providing training to staff, supervising evacuation of the building, providing first aid treatment, etc.

14.2 Other Policy Statements to include

As we have already covered these points earlier, you could also include specific plans or policies you have drawn up, or identify where the details can be found. It would be particularly useful to include references to, or outlines of, your policy on:

- smoking, and restricted areas where it is permitted or banned;
- use of drugs or alcohol on premises;
- lone workers;
- bullying or harassment in the workplace;

- staff training plans;
- emergency plans.

If possible, you should also consider how you will deal with people who are absent for long periods of time, due to illness or injury caused by work, and how you can help them get back into work as they recover – that is, a rehabilitation policy. In small firms this is not easy, but there is a lot of pressure now from governments and the insurance industry to make employers consider this question, and to find ways for them to take on more of the costs associated with long-term illness or injury.

Do not forget that you also have responsibility for outworkers or homeworkers employed by you, so they should also be included in your process for identifying potential hazards and risks, and appropriate control measures discussed with them.

In the same way that producing a business plan is not a one-off activity, the management of health and safety and other risks is an ongoing process. Your policy will, therefore, change over time as the business itself changes, and certainly the results of risk assessments will need to be reviewed and up-dated regularly. The way to ensure that the management of risks is carried out effectively in your business is to make sure that the commitment at the very top of the firm is real, and that health and safety is treated as an integral part of how you run the business. It is not just about what the law requires, but should be treated in the same way as other management issues, such as financial management and marketing. It should be very clear by now just how much it actually does influence those other issues.

In many other guides or publications about health and safety, you will often find setting the policy as the first stage in the process. By following the approach in this guide, it should be much easier to put together a realistic policy for your particular business, based on the findings of the comprehensive review you have carried out so far.

Chapter 15

Conclusion

Right at the beginning of this guide, we considered various reasons why you might have chosen to do something about health and safety in your business at this time. Having worked all the way through the guide, and hopefully completed all the activities along the way, it is worth revisiting these different motivations to see whether the guide has given you what you were looking for.

15.1 Have you achieved what you wanted from this book?

Will the inspector be satisfied?

Looking at the 'big picture' rather than separate parts of it piecemeal means you need to:

- identify current position and problems;
- deal with them appropriately;
- keep records of actions taken;
- keep control of the situation in the future.

Has working through this guide helped you to do these things? In Chapters 4–6, 7–9 and 10–12 you have identified the potential hazards and risks to people working or visiting on-site, and assessed whether existing controls are adequate. Any problems or gaps in levels of control have been identified, and a plan of action drawn up to put things right. The completed checklists, and your evidence file show what actions you have taken so far, and as targets on the action plan are met, you will be able to sign them off as completed. The work you did for Chapters 13 and 14 shows how you will continue to stay in control of the situation in the future.

Provided you carry out the actions you have identified as necessary to safeguard people now and in the future, you should satisfy the four criteria above sufficiently to demonstrate to an inspector that you are, indeed, in control of the situation. More importantly, it will have established a management approach to health, safety and fire risks that can be used to manage other risks in the business. If you had no system in place before, then this is a considerable step forward on a business and legal basis.

The accident or near-miss

In this case, working through the guide should have helped you to take an objective view of how and why an accident occurred or was narrowly avoided. It is often very difficult to look at an isolated incident in the broader context of what happens in the rest of the firm, and all too easy to blame an individual without looking critically at the details.

The activities in the earliest Chapters 1–3 will have helped to identify process or activity centres around the site, and movements of goods or people through the business. Later chapters on identifying hazards, risks and controls will also have added to the overall picture, and should have thrown some light on how and why an accident did occur. Ensuring it doesn't happen again is the critical intention behind the plan of action and the range of management controls included in Chapters 13 and 14. The actions you have taken so far demonstrate to all parties concerned that you are taking the situation seriously.

Is the insurance broker satisfied?

The use of premises and site plans as the basis of working through the guide should clearly demonstrate that you know what is happening around the site, have identified potential problems that need to be dealt with, and are managing the risks. Although a fairly basic and straightforward way to look at the business, the use of actual site plans to highlight problem areas makes it easier to explain the findings to others, but also ensures that you do actually look at your surroundings more closely. Crucially, by using the same outline to consider safety, health and fire risks it shows a much more comprehensive and co-ordinated approach than might otherwise be the case. Your evidence file should certainly show your insurers that you are managing the risks sensibly.

Will the client be satisfied?

All the elements covered in this guide should be sufficient to satisfy the requirements of major clients, particularly the risk assessment approach used. The fact that you have used the guide to produce your own evidence of how you manage risks is a positive message in itself, and the comprehensive range of statements made in your policy section should reinforce this view. The contract tender requirements always include reference to your policy on health and safety, and they inevitably want to see written confirmation that

(a) you actually know what they are talking about, and
(b) you are taking appropriate action.

The issue of outside bodies awarding certification to confirm you have a formal management system in place is a more complex one, as it depends on your own organization and preferred approach. However, the main principles of such a system are those of carrying out an audit or review of the current situation, planning for action, organizing and managing the action, keeping appropriate records, and crucially taking a continuous improvement approach. Although the evidence you have produced here may not be sufficient in itself to translate directly into a third party certified Management System Standard, it does provide an excellent base to start the process from.

Will employees or workers be satisfied?

It should be clear to everyone in the firm by now that positive steps are being taken, and indeed people will already have been involved in identifying hazards, assessing risks, and identifying where controls are inadequate. The plan of action includes reference to individuals in the firm, and the development of the policy in Chapter 14 should reinforce this.

Provided that necessary actions required are actually carried out, then workers should certainly be satisfied that their concerns have been listened to and something is being done to put things right. In particular, the prioritizing of actions on the action plan should also illustrate why everything cannot be done at once. In addition, you can also demonstrate your commitment to others outside the firm, such as those people with responsibility for placing trainees for work experience in suitable firms.

Business benefits from taking action

By now, it should be clear to you and others in the business that there are potential benefits from managing risks effectively, whether they are health, safety or fire risks. Apart from the fact that you are legally obliged to take many of the actions identified in this guide, the losses to individuals and businesses from mismanaging such risks are tremendous. Whether it is to secure future business, reduce sickness absence or high staff turnover costs, or to keep insurance premiums down, there should be real benefits to you and your business from working through this guide and managing your risks more systematically.

15.2 Relevance to your business

The fundamental approach you have taken throughout this guide should be relevant to other risks within your business. In particular, identifying hazards, assessing risks and controls should be an integral part of any business planning activity, whether it is related to expanding the physical capacity of the firm or entering new markets. Though not in any formal, glossy format, your evidence file is a valuable resource for you and others, and should certainly be a working file rather than an archive!

You may still need to gather together specific details about certain hazards or risks in the business, and Chapter 17 will give you some further sources of advice and information to contact. There are some excellent free publications from HSE, and a wide range of other sources that can give you advice, guidance or support, according to your needs. Hopefully, by now you are in a much stronger position to know what questions to ask than you might have been previously.

Part F

Industry-specific Hazards and Risks

Chapter 16

Case study examples

This chapter is broken down into individual case study examples that highlight some of the specific hazards and risks you should look out for. Many of these are common to several industries, but are still worth noting separately. It is not an exhaustive list, of course, and you should seek further guidance if there are specific things you are worried about in your own company.

16.1 Office based businesses

Includes solicitors and accountants, or administration centres of other businesses. Generally considered to be 'low risk' environments.

Safety

- Storage facilities for files, stationery or other materials – overcrowding; inadequate for weight supported; inaccessible corners, cupboards; piled too high so danger of items falling; too high up, especially for heavy boxes or files, so people tempted to stand on chairs etc. to reach.
- Minor cuts from paper edges, use of scissors, using wrong tools to open boxes or parcels; injury from use of guillotines, shredders.
- Slips and trips – trailing cables and leads; drawers left open in desks and cabinets; boxes stored under and around desks; poor floor surfaces, worn carpet especially at edges; stairs obstructed or poorly lit; wet floors in lobby areas, kitchens.
- Electric shock or burns – overloaded sockets; incorrect fuses; damaged or worn cables; electrical equipment not maintained or serviced adequately; kettles on sink draining boards!
- Burns from equipment, such as photocopier when open.
- Manual handling injuries, from lifting/pulling/pushing heavy or awkwardly shaped items
- Violence to staff in firms dealing with sensitive or confidential information.

Figure 16.1 Electrical safety is often overlooked, especially in office environments where this tangled mass of leads, plugs and sockets is not uncommon. Although it is hidden behind a desk so may not cause someone to trip, overloaded sockets and the potential for an electrical fire to start is likely then to be forgotten.

Health

- Damage from smoking – designated areas needed.
- Use of visual display units (VDUs) – regular eye tests needed for habitual users; posture and RSI problems from continuous use of keyboards; need adjustable workstation to fit the individual user; must have short break from VDU every half an hour (this does NOT mean they have to have a coffee break, but need to vary the work they do so that they are not looking directly at the screen for long uninterrupted periods).
- Lighting – adequate for the type of work being done, to reduce eye strain and poor posture, and reduce reflections or flicker.
- Heat – comfortable heat to work in (min 16°C).
- Ventilation – adequate circulation of clean air; potential hazard if air conditioning via water-cooling tower.
- Hazardous substances such as inks and solvents; cleaning materials.
- Stress – especially if threat of violence, heavy workloads, tight deadlines, insufficient staff numbers, inadequate rest breaks and facilities.

Figure 16.2 Health risks are less easy to see than safety or fire risks, so usually need a little more effort to identify. Use each work station to consider heating levels, seating and posture, general and local lighting needs, with storage at appropriate heights and reaching or twisting during work activity kept to a minimum.

Fire

- Smoking and discarded cigarettes or matches.
- Electrical faults; overloaded sockets; poor maintenance of portable electrical equipment.
- Ignition sources such as boilers, pilot lights, use of cookers.
- Fuel sources such as paper (especially when shredded and less densely packed), solvents, etc.

Security

- Controlled access to premises.
- Access to confidential records, whether paper or IT based.
- Disposal of sensitive records or material – use of paper shredder.

16.2 Small retail premises

Safety

- Delivery of goods – access doors for deliveries; where are goods stored until they can be put away properly? Shelving and racking systems adequate and strong enough; height goods are stored at; do goods cause an obstruction when delivered; use of trucks or trolleys to move goods.

- Opening boxes and containers – cuts and injuries likely from use or misuse of incorrect tools; hazard of plastic band ties and injuries to hands.
- Slips and trips – from discarded packaging; trailing cables and leads; poor, uneven floor surfaces; wet floors in main shop and storage areas; aisles and passages blocked with obstructions; steps and stairs.
- Falling items – from shelves; collapse of racks or shelves; goods stacked too high or incorrectly.
- Injuries from use of machines, such as display goods, compacters, shredders.
- Electric shock from portable electrical equipment; damaged cables.
- Theft and violence to staff – especially vulnerable lone workers, or those locking up/opening premises; paying money in at the bank.

Figure 16.3 Particular problems associated with smaller retail units are inadequate storage space away from customer access; shelves or racking systems overloaded; heavy or awkward products stored too high up, fire escape routes blocked by boxes of goods, plus security and health problems for staff.

Health

- Stress from threat of violence; long working hours; dealing with the public.
- Back injuries from incorrect manual handling techniques – goods too heavy or awkward to lift/move.
- Tiredness from standing all shift, without proper breaks or opportunity to sit down.
- Contamination of foods and health risks.
- Use of chemical substances for cleaning; storage of substances on sales shelves; damaged packaging on delivery or in storage.

Fire

- Smoking by staff or customers – discarded cigarettes or matches.
- Electrical faults, overloaded sockets, damaged cables.
- Stacked cardboard or other rubbish a source of fuel.
- Products themselves may be highly flammable, or act as fuel if fire starts.
- Arson, especially where rubbish or other goods stacked.
- Spread of fire likely to be rapid in crowded areas; escape routes need to be kept clear at all times; emergency procedure vital, and staff trained adequately to carry it out.

Security

- Locks to windows and doors; may require grilles or metal shutters (check with insurers).
- Shoplifting precautions; use of CCTV; proper lighting inside and outside premises.
- Theft from tills; reduce amount of cash kept in till at any time.
- Lone workers vulnerable, especially if a decoy used to distract them; issue personal alarms?
- Procedure for paying money into bank – vary the routine.

16.3 Hairdressers, beauticians and similar types of business

Safety

- Use of portable electrical equipment – danger of electric shock, especially in wet conditions; need to regularly check and maintain all equipment.
- Burns to staff and customers from hairdryers, infra-red lamps and other equipment.
- Scalds to staff and customers from hot water.

- Slips and trips – wet, slippery floors; trailing wires from hand-held equipment; hair and waste materials on floor; customers' bags, etc. on floor; steps or poor floor surfaces in premises; poor lighting.
- Children and hazards such as them pulling equipment down by cables; touching hot items such as curling tongs; causing trip hazard with toys, etc.
- Burns from use of chemicals in preparations.

Health

- Storage of chemicals – ensure preparations are stored separately (see HDS) and clearly labelled.
- Use of substances – fumes from substances can affect eyes, lungs, skin; inhalation from bleach-based products; act as sensitizers.
- Dermatitis and other skin reactions from direct contact with substances; burns from direct contact or splashes on skin or in eyes; hazard still present when discarding waste papers, cotton wool, etc.
- Back and leg strain from standing for long periods, or twisting actions.
- RSI from repetitive actions such as cutting hair.

Fire

- Electrical faults, or sparks as ignition source.
- Explosion from aerosol cans (if fire starts inform firemen where such cans are in premises); store cans away from direct sunlight, and preferably in a fire-resist cabinet.
- Infra-red lamps a source of heat/ignition.
- Chemical substances in combination can cause fire hazard; ferocity of fire increased by presence of substances.
- Alcohol in some dye preparations make them highly flammable.
- Oxidizing agents in some preparations provide the additional oxygen to keep fire going.
- Smoking of staff or customers, and discarded cigarettes or matches.

Security

- Money in tills.
- Paying money into bank, or collecting change – vary the routine.
- Records of clients' details.
- Customers' belongings.

Environment

- Disposal of waste chemicals – check levels are acceptable for some rinsed down sinks; collection of empty containers and used or waste chemicals.
- Disposal of waste materials such as papers or cotton wool which are still hazardous and/or flammable.

16.4 Textiles, dressmaking, furnishings

Safety

- Injury from use of cutting tools, hand-held manual or powered tools.
- Finishing by hand.
- Use of sewing machines, and danger of fingers or clothes being trapped; injury from moving parts; needles especially on larger industrial machines.
- Burns from ironing, pressing; scalds from steaming or other processes.
- Slips and trips – crowded conditions; work in progress at each work station; cables; materials stored incorrectly (insufficient storage facilities).
- Storage of heavy materials; danger of falling objects.
- Electric shock from use of range of equipment, including hand-held portable equipment.
- Sharp objects, pins, etc.

Health

- RSI – especially using scissors and sewing actions, particularly due to speed of working; hands, arms, upper body affected, made worse by twisting in seat.
- Eye strain, due to long periods of work activity – need for adequate lighting, breaks, proper equipment for job.
- Lighting generally, to avoid slips and trips.
- Back injuries due to incorrect manual handling, especially pulling, pushing, lifting heavy awkward loads; need for correct seating; specific needs of pregnant women.
- Fumes from some processes such as mercerising, and residual fumes from dyeing processes; causing skin/eye/throat/lung irritation or acting as sensitizer.
- Noise levels – especially as likely to be over prolonged periods, and background as well as local source.
- Stress – due to speed of work; piecework; team work and pressure to work to level of the quickest person; inadequate training; volume of work.
- Adequate ventilation vital.

Fire

- Fibres and dust in the air, danger of explosion and rapid spread if fire starts.
- Electrical equipment and sparks; proper maintenance needed.
- Heating appliances, especially portable or old and inefficient models; overcrowding of premises.
- Flammable fabrics and materials to provide fuel for a fire; particular danger of toxic fumes or smoke produced.

- Storage and usage of chemicals for different processes; cleaning materials and lubricants for machines.
- Emergency procedures in place; exits clear of obstructions at all times; fire warning systems in places little used where fire could smoulder unnoticed for some time.

Security

- Theft of items, materials, finished garments by staff.
- Theft and robbery by others.
- Opportunities for arson.
- Theft of designs and intellectual rights.

Environment

- Disposal of waste packaging; provision of appropriate packaging for own finished product.
- Disposal of waste materials.
- Disposal of hazardous waste products from processes.

16.5 Mobile services such as chiropodist, hairdresser, dog grooming

Safety

- Personal safety on other people's premises, including danger presented by people or animals.
- Injury from carrying equipment or materials; items falling.
- Injury to client from processes or materials used.
- Use of portable electrical equipment – faulty supplies or sockets; faulty or damaged equipment.
- Slips and trips – working in unsuitable surroundings, overcrowded or many obstacles; steps and stairs.

Health

- Manual handling injuries to back, carrying equipment and materials.
- Other musculoskeletal injuries due to posture during work; possible RSI.
- Hygiene and contamination through skin/ingestion/inhalation; insufficient washing facilities available.
- Use of chemicals and other hazardous substances (refer to type of work carried out).

Fire

- Storage of chemicals and other substances at home.
- Electrical faults.
- Explosion from use of aerosol cans.
- Note insurance cover adequate for equipment and materials at home or on others' premises.

Security

- Personal danger when working alone, especially outside normal working hours.
- Travelling alone.
- Transporting equipment and materials – theft or accidental loss.

Environment

- Disposal of waste products and materials.

16.6 Repairers such as jewellery, watches, dental, or shoes (also see retail section)

Safety

- Use of fine drilling or abrasive equipment.
- Flying pieces of metal or other materials in eyes.
- Danger of burns from materials or process.
- Use of small, hand-held cutting tools; need to keep them sharp and well maintained.
- Portable electrical equipment and electric shock.
- Small-scale soldering or welding.

Health

- Eyesight at risk with close work; need for adequate lighting in work areas; may need 'daylight' bulbs or measures to reduce glare.
- Posture at workbench, especially when carrying out close work over long periods of time.
- Need for adequate breaks.
- Use of hazardous substances, including solvents, adhesives, and cleaning materials; irritant or damage to skin/eyes/throat/lungs.
- Fumes produced by soldering or welding; need for adequate PPE.
- Appropriate ventilation systems needed.

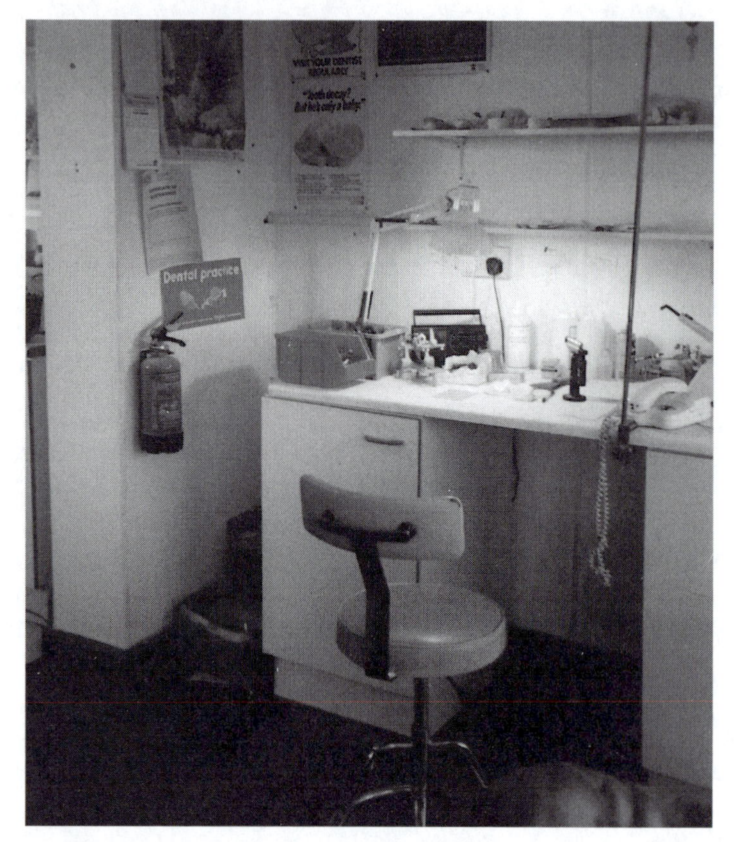

Figure 16.4 Obviously, it is much more efficient if the work station is well organized, to reduce health, safety and fire risks. If extra local lighting is needed for close or detailed work, then make sure it is adjustable and anchored securely. Some tools used for fine repair work also present burning, cutting and abrasion injury hazards.

Fire

- Chemical substances and rapid spread of fire.
- Fumes from processes; build up of fumes from open containers.
- Storage of hazardous substances.
- Flammable fine dusts and powders and danger of explosion.
- Electrical faults, heat sources, sparks as ignition source.
- Flammable materials as fuel if fire starts.

Security

- Safe care of customers' goods.
- Confidential records.
- Theft of substances or other materials.

Environment

- Disposal of hazardous waste substances; disposal of empty containers.
- Disposal of waste materials.
- Danger of arson.
- Exhaust emissions adequately controlled.

NB For shoe repairers, there will be additional safety factors associated with large pieces of machinery, such as punches, stapling machines and edge trimmers. All of these are extremely hazardous of course, so will need to be considered in the same way as those in engineering.

16.7 Crafts and pottery

Kilns

Safety

- Burns from direct contact with kiln.
- Electrical shock from faulty equipment or systems; overloading sockets.
- Trapped or crushed by moving parts when loading or unloading; falling objects.

Health

- Manual handling when pulling, pushing, lifting heavy awkward loads.
- Noise levels, especially from ventilation or exhaust systems.
- Ventilation adequate; fumes from some processes.
- Maintenance – insulation in some kilns may include asbestos or other hazardous mineral fibres so need special attention.

Fire

- Use of natural gas or LPG to fire kiln as ignition/heat source.
- Storage of fuels.
- Naked flame/pilot lights as ignition source – use of automatic ignition systems is preferable.
- Build up of gas in low level spaces around the kiln, as explosion hazard or toxic fumes.
- Wooden ceilings above kilns at risk from intense heat.

Environment

- Ventilation and exhaust systems.
- Disposal waste materials; disposal empty gas cylinders.

Pottery

Safety

- Electric shock and portable electrical equipment; insulation in wet conditions.
- Cuts, crush injuries from moving parts of machinery for example, cutters, kneaders, mixers.
- Unsafe racking and storage systems, especially in relation to height and weight of loads stored; danger of falling objects.
- Slips and trips – wet floors, especially with clay; trailing hoses and cables; materials stored on floor.

Health

- Lighting levels; eyestrain from prolonged fine or detailed work.
- Hazardous substances in clays, paints, etc. – includes lead, selenium, cobalt, silicon.
- Restriction of access for pregnant or nursing women.
- Fumes from furnaces/kilns.
- Temperature and humidity levels; need for proper ventilation.
- Use of cleaning materials, as irritants to skin and eyes, plus inhalation and lung damage.

Fire

- Electrical faults and equipment.
- Storage of materials as fuel source.
- Dusts and fumes and danger of expolsion.

Environment

- Disposal of waste materials and products.
- Disposal into domestic drainage systems.
- Ventilation and exhaust systems.

Jewellery

(This also includes many similar points to those in the pottery or textiles sections.)

Safety

- Use of metal wire, causing cuts and piercing injuries.
- Use of cutting tools, whether manual or powered.
- Small welding or soldering tools, and burns or flying pieces.

- Electric shock and use of electrical equipment; faulty; overloaded sockets.
- Storage of scrap and sharp objects, potential for cuts, piercing injuries.

Health

- RSI with repetitive twisting movements of fingers.
- Eyestrain and inadequate lighting.
- Use of paints, glues, solvents and hazards of skin contact or inhalation.

Fire

- Heat from processes as source of fire.
- Chemical substances highly flammable, danger of explosion, rapid spread of fire.
- Storage of materials as fuel source.
- Storage or use of flammable packaging materials.

Environment

- Disposal of waste materials and substances; ventilation and exhaust systems.

16.8 Dentist

Safety

- Use of portable and permanent electrical equipment; electric shock in wet conditions.
- Correct use of hypodermic syringes; danger of puncture wounds to staff and patients.
- Correct use of drugs and anaesthetic procedures.
- Personal safety and violence from uncontrolled patients, possibly drug-induced behaviour.

Health

- Contact with transmittable diseases by breathing; skin; saliva or blood samples.
- Hazards from incorrect use of drugs, either for staff or patient.
- Posture, sometimes in awkward twisting position; back strain.
- Need for good lighting at point of work.
- Use of X-ray machines; note vulnerable workers such as pregnant women or those of child-bearing age.
- Use of VDUs and computer systems (see also Office section).

Fire

- Electrical faults.
- Use and storage of chemical substances.
- Additional oxygen supplies; storage and disposal of cylinders.

Security

- Storage of and access to drugs.
- Storage of and access to patients' records.
- Potential for break-ins or robbery of drugs, syringes, etc.

Environment

- Disposal of sharp waste like syringes.
- Disposal of waste products and materials.
- Disposal of chemical waste.

16.9 Vets and animal establishments

Most of the items listed for dentists will also apply to vets, and those listed below are additional.

Safety

- potential for crush/gore/bite/sting injuries from handling animals
- need for appropriate protective clothing
- stacking materials, foodstuffs etc, and danger of falling objects
- falls off steps, ladders; falls from animals

Health

- Use and storage of chemicals or drugs, many of which are stronger for use with animals than for humans.
- Adequate eye and skin protection needed; inhalation of some preparations particularly harmful.
- Manual handling injuries with bales of hay, animal foodstuffs, animal equipment such as saddles.
- Diseases passed from animals to humans (zoonoses), tetanus, presence of pesticides.
- Use of X-ray machines and radiation.
- Access to adequate washing and cleaning facilities.

Fire

- Highly combustible foodstuffs and hay.

Environment

- Disposal of medicines, dressings, etc.
- Disposal of carcasses and animal waste.

16.10 Agriculture and horticulture

Safety

- Accidents with vehicles on site; use of trailers, bailers, harvesters.
- Rough terrain vehicles and tractors – overturning.
- Cutting or sawing equipment and large machinery – crushing/cutting/amputation hazards.
- Use of hand-held tools and equipment; electric shock from faulty equipment, supplies or voltage.
- Falling into large containers such as silos, tanks, pits; need for proper lighting, fencing, guarding, especially against access of children.
- Working in confined spaces.
- Storage areas; falling objects; collapse of stored goods and shelving.
- Use of pressure equipment.
- Use of ladders; working at heights inside and outside buildings.
- Crush/gore/bite/kick injuries when handling animals.
- Particular hazards associated with children on site, and when working alone in isolated area.
- Handling guns and ammunition.

Health

- Noise levels, especially in large animal/bird houses at feeding times; machinery or vehicle noise.
- Need for eye protection when using some cutting equipment.
- Use of chemicals causing skin or eye irritation, burns, breathing difficulties (especially in horticulture); hazards associated with use of sheep dip.
- Dusts and grains, causing asthma, lung and throat damage; note that permanent disablement is likely.
- Storage and transportation of hazardous substances and materials.
- Diseases spread from animals to humans; pregnant women should not be involved during lambing.
- Fumes from faulty gas appliances.
- Manual handling injuries – dealing with animals, harvesting or picking crops, moving foodstuffs.
- Handling and storage of veterinary medicines.
- Stress – working hours; dealing with extreme weather conditions; economic environment.

Fire

- LPG and oil fuel storage and use.
- Explosion hazard from dusts and grains.

Figure 16.5 CD: Hazard data sheets must identify the potential harm from misuse or storage of substances, so the use of chemicals for crop spraying, for instance, must take these details into account to safeguard the health of workers, passers by and consumers.

- Storage of hazardous substances, chemicals, fertilizers; storage of petrol for vehicles.
- Highly combustible materials like hay stacked in large quantities, so rapid spread of fire.
- Potential fire sources often close to living accommodation.
- Ammunition.
- Rubbish and other waste materials.

Security

- Opportunities for arson, given the points raised above.
- Protection of public from animals, especially with public rights of way on land; control to avoid escape of animals.
- Fencing or other means to secure large areas of land from trespass.
- Secure storage, and use of, guns and amunition.

Environment

- Water extraction sources; land drainage.
- Contamination of land or water courses.
- Seepage from slurry tanks or pits; control of septic tanks and emptying.

- Opportunities for water, wind and sun generation of power.
- Leakage from oil storage tanks; disposal of used oil or petrol products.
- Hedge and woodland management and disposal of scrap materials.
- Use of controlled burning.
- Use and storage of pesticides, fertilizers, other hazardous substances.
- Disposal of waste materials and chemicals; disposal of medicines, empty containers.
- Disposal of redundant equipment, vehicles or machinery.
- Disposal of animal waste and carcasses.

16.11 Florist

Safety

- Use of wire cutting tools; heavy duty cutting tools for plant materials.
- Scratches, cuts, and puncture injuries from plant materials.
- Slips and trips – water spillages; obstructions such as flower bins in aisles or work areas; trailing wires; poor floor surfaces, overcrowded areas. Consider customers on premises.
- Electric shock from use of electrical equipment in wet conditions; hazards from localized heating appliances.
- Buying and delivery of large bulky supplies – handling injuries; storage facilities.
- Driving hazards when making deliveries to customers.

Health

- Tetanus and other diseases from handling plants.
- Handling and storage of hazardous chemicals for care of plants; potential for allergy sensitizers.
- RSI, with small, twisting repetitive movements; posture and standing for long periods.
- Temperature and humidity, especially working in cool conditions.
- Stress – contract work, emergency work for funerals, for example; weddings; dealing with and advising customers.

Fire

- Combustible materials such as dried foliage, packaging materials.

Environment

- Disposal of plant materials and waste.

16.12 Forestry

Safety

- Use of chain saws – maintenance; training; guards; sharpened; brakes; correct PPE including helmet/ear defenders/goggles/gloves/leg protection/boots/no loose clothing. NOT to be used when working alone.
- Use of circular and other saws – using push sticks; proper guards in place; grippers; blades sharpened; speed set correctly; working height set properly.
- Shredding machines – correct guards in place and correct use of push sticks.
- Ladders, lifting gear in good working order and regularly checked; use of harnesses.
- Hazards of overhead electricity cables; electric shock from faulty or damaged electrical equipment; incorrect voltages.
- Use of barriers and warning signs to protect others.

Figure 16.6 Some tasks, such as tree felling are inherently hazardous, and it is virtually impossible to make them safe. In this case, safe procedures and correct use of equipment/PPE is essential. In this photograph, proper harnesses, clothes, boots, goggles, hat and ear defenders are used, and comprehensive training received by the worker. There is also appropriate support from others at ground level.

- Escape routes identified beforehand; pre-plan risk zones, ground conditions, wind direction, etc.
- Use of hand tools for digging, planting.
- Loading wood onto vehicles – securing the weight correctly, ground conditions suitable for vehicle when loading; dangers associated with transporting load.

Health

- Manual handling of heavy, awkward loads, and danger of back injuries.
- Noise from use of machines.
- Dusts – especially highly carcinogenic hard wood dusts.
- Use of or contact with wood treatments, preservatives, other hazardous chemicals.
- Use of or contact with pesticides, fungicides, insecticides.

Fire

- Sparks from use of electrical equipment.
- Highly combustible materials, with potential for rapid spread with small shavings etc. and intensity of fire from stacked materials.
- Risk of fire or explosion from use and storage of chemical substances.
- Risk of fire or explosion from use and storage of oil and petrol products used in vehicles.

Security

- Make area secure from unauthorised entry when working.
- Also ensure area secure if left unattended.
- Safekeeping of chain saw and other hazardous equipment.
- Vehicle security.
- Theft of stored wood during and after work completed.
- Risk of arson.

Environment

- Disposal of unwanted wood and other plant materials.
- Transporting wood and scrap materials.

16.13 Carpenter and joiner, and furniture manufacturers

Safety

- Slips and trips – need for good housekeeping standards; hazards from off-cuts of materials; trailing cable, leads, ventilation or compressor hoses; wet floors from water, adhesives, solvents, etc.

Figure 16.7 An excellent example of a tidy workshop area, with appropriate storage that safeguards the tools as well as the person using them. Information on the correct use of the equipment is at hand, plus records of maintenance checks carried out.

- Use of powered tools; machines with moving parts – correct guards in place and use of push sticks.
- Cuts from saws, circular or band saws, other cutting tools, especially kickbacks.
- Electric shock from faulty or damaged electrical equipment; overloaded sockets; incorrect voltage for location; use of circuit breakers vital.
- Hazard from materials falling when incorrectly stored or stacked; collapse of wood stores or racking systems.
- Use of ladders, scaffolding, steps – regular checks, wear harnesses as appropriate, check conditions if working on someone else's site (see Window cleaner and Roofing sections).

Health

- Noise from use of some electrical or powered tools.
- Lighting adequate at point of work, especially if portable lighting needed.
- Manual handling injuries to back and upper body from working in awkward positions for long periods; carrying, lifting, pulling and pushing large, heavy or awkward loads.
- Dust hazards from cutting, sawing, drilling – lung/skin/throat/eye damage especially cancer from some wood dusts.

- Need for local exhaust systems at point of work.
- Use and storage of hazardous substances such as glues, wood treatments, smokes, solvents, paints; especially note potential build up of fumes from open containers.
- Working in confined spaces such as floor or loft cavities – ventilation and emergency escape systems.
- Protection of skin from harmful rays in direct sunshine; potential for skin burns.

Fire

- Electric sparks from use of electrical equipment; faulty equipment or overloading sockets.
- Explosion from concentration of wood dusts, shavings, etc.
- Use and storage of flammable substances (glues, treatments, paints, solvents, etc.).
- Use of compressed air equipment; gas or oxygen cylinders as additional oxygen sources.
- Transporting hazardous substances – amounts and type of containers used.
- Combustible furnishing fabrics and materials, especially foam based fillings; storage of finished goods; packaging materials.
- Waste rags or materials containing traces of flammable substances.

Security

- Securing premises when left unattended.
- Safeguarding equipment and tools when transporting or working.
- Potential for arson, especially in waste skips.

Environment

- Disposal of chemical waste, and empty containers.
- Disposal of scrap wood and other materials.
- Ventilation and exhaust systems.

16.14 Roofing

Safety

- Use of ladders and scaffolding – potential for falls; use of harnesses; barriers and safety net systems (see also Window cleaner).
- Correct checking, maintenance and use of hoists and other lifting gear; falling objects.
- Cuts and injuries from damaged roof surface, tiles, rafters.
- Falls through skylights, fragile roofs, hatches; sloping roofs.

- Electric shock from use of faulty or damaged electrical equipment; use of power tools; especially wet weather conditions.
- Overhead power lines.
- Burns from use of bitumen type products, heat torches.

Health

- Heat-solvent chemical agents in bitumen products, affecting skin and breathing.
- Contact with mineral fibres – glass wool; insulation materials; potential lung damage or skin irritation.
- Asbestos, especially in tiles, lagging; exposure can cause cancer.
- Dust hazards when cutting tiles.
- Excessive heat, cold, wet conditions.
- Manual handling injuries from lifting, carrying, working in awkward positions.

Fire

- Working with flammable materials, including chemicals, wood, thatch.
- Smoking as source of ignition.
- Use of direct heat sources, such as with bitumen products.
- Explosion – solvents in enclosed spaces; concentrations of dust or wood shavings.

16.15 Window cleaner

Safety

- Use of ladders – regular checking for signs of damage; maintenance; proper height for job; securely footed, and fixed if above first floor level; danger of side reach.
- Hazards to passers by of falling objects; use of barriers or warning signs.
- Proper use of safety harnesses, with secure fixing to safety eyebolts.
- Falls from height, off ladders, from cradles, and from ledges or sills, etc. – adequate training in procedures and use of safety devices.
- Use of roof boards – two at a time (one to use and one to move).
- Work from scaffolding – check safety, especially if on someone else's site.
- Cuts from broken glass, through glass.
- Need to check windows, sashes, locks etc. before starting to clean.

Health

- Use of chemicals and cleaning fluids – skin damage or breathing difficulties.
- Damage to skin from ultra violet rays when exposed directly to sun.

Security

- Storage of ladders; safeguarding access to ladders when left unattended.

Environment

- Disposal of cleaning substances.

16.16 Construction (general points)

The major hazards apply to most trades on construction sites, generally in the following categories:

- working at heights;
- falls from or through roofs, windows, floors, stair shafts, unguarded areas;
- electric shock, and use of portable electrical equipment; overhead cables or trailing leads;
- injuries from moving parts when using equipment, machinery and tools;

Figure 16.8 Very specific requirements are laid down for working in the construction industry, particularly for the erection of scaffolding and use/maintenance of lifting gear, for instance. As a small contractor on a larger site, you must comply with the health and safety requirements set down by the main contractor, and you also have a responsibility to make sure anyone else you sub-contract work to complies with these requirements.

- burns from use of blow torches or other heat sources;
- pressurised containers and equipment and risks of explosion;
- working in confined spaces or underground workings;
- manual handling – lifting, carrying, pulling, pushing large/heavy/awkward loads;
- effects of heat, cold, humidity, wet (including skin cancer from exposure to the sun);
- use of hazardous substances – storage, handling, disposal;
- vehicle safety on site, including potential for overturning.

In addition, there are specific regulations that you need to comply with even if you only employ two or three workers. These are the Construction Design and Management Regulations (CDM), which apply at every level of the construction project, from client/designer/planning supervisor/principal contractor/and contractors or self employed individuals.

16.17 Butchers

Safety

- Use of knives, handsaws, cleavers – need to be well maintained, sharp, stored correctly; appropriate PPE must be worn, including chain mail apron and gloves where necessary; young people especially vulnerable.
- Use of cutting machines such as mincers, mixers, saws, grinders – need proper maintenance and training in use; ensure guards, locking devices, push sticks used correctly at all times.
- Slips and trips – wet floors (non-slip surfaces possible?), grease and oil products; trailing cables.
- Electric shock from faulty or damaged electrical equipment, and when used in wet conditions.
- Some machines are 'prescribed dangerous machines' so specified steps must be taken to protect people – check with HSE.
- Need appropriate lighting to reduce potential for accidents; use of circuit breakers.

Health

- Contamination from meat and other foodstuffs (especially reference to BSE contamination).
- Manual handling injuries from heavy weights; standing for long periods; upper body damage from twisting and cutting actions.
- Use and storage of cleaning substances.
- Temperature effects, especially working in cool temperatures, or in chill rooms – slowing down of reactions and manual dexterity.
- Separation of cooked and raw meat products.

Figure 16.9 As well as health risks to employees working with meat and other food products, food hygiene controls are very stringent and are in addition to health and safety requirements. Many Local Authorities have produced their own guidance for food outlets which reflect the local conditions, so contact your local Environmental Health Office for details.

Fire

- Faulty electrical equipment; overloading sockets.

Security

- Safeguarding money in tills; paying in to bank; theft.
- Procedures for walk-in freezer or chill cabinets.

Environment

- Disposal of waste materials and meat products.
- Disposal of chemical or hazardous substance waste.

16.18 Catering

Safety

- Injury from machinery moving parts, grinders, mixers, mincers, washing up machines, rotating-table machines (see also Butchers section).
- Use of knives, cleavers other sharp implements.
- Burns and scalds – hot surfaces, liquids, direct heat sources in ovens and grills.
- Steam from dishwashing machines, kettles, food from microwave ovens.
- Slips and trips – wet and greasy floors; obstructions; trailing cables and hoses; storage.
- Falls – floor surfaces, steps and stairs, layout of work and customer areas, carrying heavy items, poor lighting; customers' bags, coats etc. near tables and service area.
- Electric shock – faulty electrical equipment, insufficient maintenance, use of portable appliances; especially in wet conditions.
- Falling objects from shelves, tables; includes large equipment not anchored correctly.

Figure 16.10 In a small snack bar environment, such as this one, space is always at a premium. This is particularly hazardous if more than one person is working at a time as slips, trips and accidents using kettles, knives or hot equipment are more likely.

- Cuts from broken glass, crockery, opened cans.
- Use of compacters, waste disposal units.
- Customer safety, especially when serving food (i.e. not over their heads).

Health

- Use of cleaning substances – skin irritants, eyes, breathing difficulties, burns.
- Food contamination during preparation; airborne contaminants.
- Standing for long periods; silver service action of heavy weight on extended arm; pulling/pushing stacked trolleys of food or crockery.
- Adequate lighting required at workstations to reduce eyestrain; posture when carrying out close presentation or preparation work.
- Heat, cold, humidity and need for proper ventilation and exhaust systems.
- Leakage of gas fuels.
- Stress – working conditions, long hours, breaks, dealing with the public and possibly complaints.

Fire

- Electrical faults; sparks from use of electrical equipment.
- Burning fat and grease – spills, overheated, old deposits of particles.
- Other ingredients flaring during cooking under grill or on hob.
- Use and storage of LPG and other gas fuels, risk of explosion.
- Steamers or boilers and risk of explosion.
- Use and storage of aerosol cans.
- Smoking and discarded cigarettes or matches.

Security

- Safeguard against deliberate food contamination.
- Personal and company money or other valuables.
- Potential for theft or pilfering.
- Unauthorized access to premises.
- Theft or damage to customers' valuables and belongings.

Environment

- Water usage.
- Disposal of substances and materials down sinks; disposal empty containers and gas cylinders.
- Waste disposal of food or other items.
- Pest control.
- Exhaust ventilation systems.

16.19 Hotels, guest houses, etc.

Note it is particularly important to consider these issues in relation to staff, contractors, entertainers, casual workers and guests.

Safety

- Kitchen areas – see Catering section above.
- Restaurant – also see Catering section above.
- Bedrooms: injury when lifting or moving furniture; sharp corners or edges of furniture.
- Use of portable electrical equipment such as vacuum cleaners; faulty or damaged equipment; guests' own appliances.
- Bar areas – see Pubs and Bars section.
- Office and reception areas – see Office section.
- Grounds: injuries from vehicles on site, including guests' and delivery of goods.
- Ground maintenance such as grass cutting – electrical equipment and need for circuit breakers; flying debris; trailing wires; broken glass or other sharp objects.
- Maintenance of buildings – see also use of ladders and equipment in Window Cleaner and Carpenter sections.

Health

- Manual handling injuries from moving heavy or awkward loads.
- Use of cleaning and other hazardous substances; effects on skin/eyes/breathing/digestive system; danger of mixing chemicals; use of pesticides or other preparations.
- Potential for Legionnaires Disease in water storage tanks/cooling systems.
- Smoking, especially for staff in areas where the public is allowed to smoke; appropriate and adequate ventilation systems required.

Fire

- Smoking of staff or customers, with discarded cigarettes or matches.
- Combustible materials as fuel, with soft furnishings, bedding, table linen; warning systems in storerooms.
- Heating systems and potential for explosion.
- Faulty or damaged electrical equipment; overloaded sockets; overloading with infrequent events causing extra drain on power, such as entertainment/shows/discos.
- Smouldering debris and bonfires in grounds.
- Potential for arson.

Security

- Theft of money, valuables, goods from guest rooms.
- Safeguarding valuables of guests.
- Fraud (staff, guests or others).
- Vehicle security; adequate lighting inside and outside buildings, especially in isolated areas of grounds.
- Information records of guests.
- Key control and signing in of guests; car park controls.

Environment

- Kitchen waste disposal.
- Chemical and hazardous substances disposal.
- Noise levels at events.
- Water usage; spillages and contamination of ground or water courses.
- Exhaust and ventilation emissions.

16.20 Pubs and bars

Safety

- Crush injuries from handling barrels, kegs, crates, other containers.
- Working in confined spaces – cellars, especially danger of leaked carbon dioxide from cylinders.
- Burns from frosted cylinders; scalds from glass washing machines when opened.
- Cuts from broken glass, bottles, other containers.
- Electric shock – use of portable electrical equipment; faulty or damaged equipment and cables; overloaded sockets; especially with wet hands, and where floors are wet.
- Slips and trips – trailing cables; obstructions in passageways and storage areas; inadequate storage facilities and space; wet or greasy floors; poor floor surfaces, steps and stairs; inadequate lighting or guarding of open areas; falls into cellars and other hatch openings.
- Use of steps and ladders.
- Violence – major concern especially late night; when staff leave or enter premises; paying in to bank; avoid workers being left alone in bar areas with customers; violence can be from other staff as well as customers. System required for dealing with violence and notifying the Police or emergency services.

Health

- Smoking – whether as individuals or as the effects from other people smoking; ventilation systems vital, and properly considered smoking policy.

Figure 16.11 Storage in this service area at a vineyard shop is well planned and laid out, uncluttered and attractive. Heavy crates of wine bottles are stored elsewhere at low levels, to make transporting to customers' cars safer for everyone.

- Manual handling injuries from lifting and carrying.
- Use of hazardous substances, including cleaning materials, pipe cleaning fluids, ammonia based products.
- Noise levels in bars, whether taped or live music.
- Hypodermic syringes of customers, and potential contamination.

Fire

- LPG cylinders, kegs and pressurized vessels or systems; danger of explosion.
- Alcohol products as fuel leading to rapid spread of fire.
- Faulty electrical equipment; overloading sockets.
- Open fires, burning wood or coal, as source of sparks and ignition.
- Smoking and discarded cigarettes or matches.
- Room heaters; potential for portable heaters to be overturned.
- Oil fired boilers and systems.
- Rubbish, waste paper and other scrap materials piled up; potential for arson.

Security

- Vandalism or arson, to premises or car parks, to business or customer property.
- Theft and robbery; fraud; theft by staff.
- Bomb threats.
- Banned individuals; under-age drinkers; drug abuse on the premises; gambling.
- Safeguarding customers' belongings.
- Vary routines and staff involved in paying money in to bank.
- Controlled access to premises and specified areas inside building.

Environment

- Disposal of used cooking oil and other hazardous substances.
- Disposal of broken glass; drums and containers; empty cylinders; cardboard and other packaging.
- Safe disposal of medical waste and syringes if found.
- Noise levels at events and when customers leave premises at closing time.

16.21 Vehicle repairs

Safety

- Working in pit areas, confined spaces; falling in to pit (note these pit areas will be phased out in favour of lifts that raise the vehicle itself).
- Insecure hoists, jacks and other lifting gear – need for regular checks and maintenance; falling objects.
- Use of pressure equipment, radiator pressure caps, etc.
- Portable electrical equipment – faulty or damaged equipment; overloading sockets; incorrect fuses used; danger of shock, especially in wet conditions.
- Slips and trips – trailing cables and hoses; damaged floor surfaces; debris lying around floor and work areas; spilt oil and grease, especially when combined with water.
- Potential injury from use of machines and tools, including grinding/ sanding/cutting; moving parts in engines when exposed.
- Customer safety on site; hazards from vehicles being driven on site, especially if drivers unfamiliar with site/type or size of vehicle/ controls.
- Contact with acids and alkalis – brake testing, battery acid, etc. Note they can burn through ordinary protective clothing so need to have the correct gloves and other PPE.
- Contact with adhesives, sealers, causing severe skin damage.
- Working on air-bag units from vehicles as they contain gas generators.

Figure 16.12 A typical example of poor housekeeping in a workshop, making an accident or incident inevitable. Trailing cables and airlines compete with other obstacles on the floor; unsafe racking is overloaded; various flammable materials sit next to ignition sources; and corrosive substances are not stored correctly.

Health

- Some fumes can be heavier than air, so likely to settle in lower sheltered levels especially pits.
- Exhaust fumes from vehicles.
- Potential for skin cancers from use of grease and mineral oils; use of adhesives and solvents leading to dermatitis or similar types of harm; or acting as sensitizers.
- Breathing fumes from welding and other processes.
- Manual handling injuries from working in awkward, crouched conditions; lifting and pushing heavy loads.
- Noise from use of some tools and processes.

Figure 16.13 Washing facilities must be adequate for the type of work being carried out. As oils and grease are carcinogenic (with the potential to cause cancer), it is vital that correct cleaning agents are used, with a good supply of hot water and hand drying facilities available.

- Eye protection needed from flying debris, falling bits of rust, etc. when working underneath vehicles, splashes of chemicals; need for adequate light at point of work.
- Radiation damage from welding or use of lasers.
- Asbestos in brake and clutch linings and pads – OK in normal usage, but exposure when drilling, grinding or filing them.
- Need for adequate washing facilities, correct cleaning preparations, and barrier creams.

Fire

- Use and storage of flammable substances, solvents, paints adhesives, etc.
- Smoking and discarded cigarettes or matches; explosion hazard if residual fumes on clothing or hands after certain processes.

- Oil rags potential source of ignition – must be in fire-resist containers, and only kept in small quantities.
- Sparks from use of electrical equipment, welding.
- Explosion risk from gas bottles and cylinders, LPG, pressure equipment.
- Petrol and oil as fire risk. NOTE petrol must not be syphoned by mouth or used as cleaning agent!

Security

- Theft or damage to tools or equipment.
- Vandalism or arson; restricted access to parts of site, especially for customers.
- Safeguarding customers' property.

Environment

- Disposal of oils, solvents, other hazardous substances; leakage of oil tanks into ground.
- Separation of waste products; disposal of large scrap and electrical waste properly.
- Noise pollution, especially some heavy processes.

16.22 Engineering (see also Vehicle repairs section)

Safety

- Injury from moving parts of equipment and machinery – unguarded or disabled safety features; old and worn machinery, leading to noisy running and attempts at short-term fixing instead of proper maintenance.
- Use of pressure vessels and steam plant.
- Use of hoists and other lifting gear; hazards of falling objects, or swinging beams or loads.
- Correct use of fork lift trucks essential – proper training, controlled access; blocked aisles; clearly marked pedestrian routes; use of trolleys and other means to move heavy items around workplace.
- Slips and trips – trailing cables, leads, air hoses, obstructions, poor floor surfaces and inadequate lighting.
- Falls from heights, ladders, stairs, open shaft or pit areas unfenced; proper use of ladders (see Window cleaner section).
- Working in confined spaces, such as inside tanks or tanker bodies, vehicle pits, etc. – ventilation and escape routes/procedures vital; correct PPE to be worn, and rescue harnesses.
- Electric shock from faulty or damaged electrical equipment; overloading sockets; no circuit breakers in use.
- Hand/arm vibration damage from use of some tools or equipment.
- Burns, eye damage or piercing from flying particles during welding process.

Figure 16.14 This fork-lift truck is clearly not loaded safely! Items are not secured and can easily fall when the truck moves and turns corners, they are too high so that they obstruct the driver's view, and the load is likely to be too heavy for this particular truck. Adequate training and supervision to ensure drivers follow correct procedures is vital.

Health

- Exhaust fumes from vehicles moving on site, including some lift trucks.
- Noise from machines in use; general noise levels; correct use of PPE hearing protection.
- Extremes of heat or cold; need for adequate ventilation and exhaust systems in different parts of work area; must be appropriate and properly maintained.
- RSI and repetitive twisting movements with some tasks; manual handling injuries from lifting, pushing or pulling big, heavy or awkward loads.
- Fumes or dusts from processes; radiation hazards, and need to identify particularly vulnerable groups of people that need protection.
- Exposure to substances that can damage eyes/throat/skin/lungs/digestive system, or act as sensitizers, such as acids and alkalis; adhesives; solvents and degreasing agents; mineral oils; carbon dioxide poisoning.

- Need for adequate washing and rest facilities, with correct cleaning substances and barrier creams.
- Stress – heavy workload, shift patterns, insufficient breaks, high noise levels.

Fire

- Potential ignition sources from sparks; static electricity; friction; oxyacetylene and welding sparks.
- Electrical faults; overloaded sockets; damaged cables.
- Explosion risk from pressure vessels or pressurized containers, including aerosol sprays.
- Explosion and rapid spread of fire from storage or use of chemical substances.
- Dust as explosion risk; poor ventilation.
- Broken or disused pallets stored next to building; risk of arson.

Security

- Access restricted to site, buildings, stores of hazardous substances.
- Access to equipment and materials, including air lines.
- Movement and security of vehicles on site.

Environment

- Separation and disposal of waste materials, substances; disposal of obsolete machinery.
- Exhaust emissions; seepage of oils or other substances into ground or water course.
- Noise levels.
- Choice of suppliers to ensure environmentally friendly materials where possible.

16.23 Petrol filling stations and car wash facilities

Safety

- Shop area – see Retail section, but also note hazard of large expanses of glass, glass doors, etc. and need to make them obvious to customers; edges and corners of shelf units safe.
- Vehicle maintenance – see Vehicle repair section, but add need to provide additional advice and guidance to customers.
- Forecourt area – slips and trips with oil, grease and water on floor; mud, ice and water additional hazard during winter – are procedures in place to reduce this risk to customers and staff?

- Hazards from vehicles moving on site; particular hazard when fuel deliveries taking place, with restricted vision, reversing vehicles, drivers unfamiliar with site layout.
- Electric shock from use of electrical equipment, especially in wet conditions.
- Car wash area – electrical safety in use and during maintenance.
- Injury from moving parts; air compressor units.
- Procedures for assisting customers in event of breakdown or malfunction to ensure their safety.
- Slips and trips – oil/grease/water/detergents/waxes; trailing cables and hoses, when car wash stationary or moving.

Health

- Exposure to hazardous substances such as detergents, oils and waxes.
- Fumes from petrol.

Fire

- NO SMOKING
- Sources of ignition, such as engines running; mobile phones need to be switched off in forecourt area.
- Electrical faults; faulty or damaged electrical equipment; power failure; insulation for use in wet conditions.
- Use of air lines – potential for injury and explosion; additional oxygen supplies.
- Explosion risk from large quantities of petrol stored; petrol fumes in atmosphere.
- Additional hazards where goods are sold and stacked on shelves together, such as wood and barbecue charcoal next to lighter fuels, next to firelighters, next to other ignition sources such as gas cylinders!

Security

- Theft of petrol; use of CCTV and other security measures; ensure adequate lighting inside and outside premises.
- Robbery, especially late or early opening stations; ensure adequate protection for employees, and proper training and procedures to deal with such an event.
- Shoplifting and theft, especially cigarettes and high value goods.
- Premises and valuables can be secured if staff have to assist customer.
- Adequate protection for lone workers.
- Emergency procedures must be in place to deal with fire/bomb threats/robbery/power failure/accidents to staff or customers.

Environment

- Correct storage of petrol to avoid leakage and contamination of surrounding areas.
- Disposal of water with waste cleaning products from car wash facilities.
- Dispersal of exhaust and other fumes effective.

Chapter 17

Sources of advice and guidance

17.1 Government departments

Health and Safety Executive

HSE Information Centre
Broad Lane Sheffield S3 7HQ

Tel: 0114 289 2345 fax: 0114 289 2333

HSE Books
PO Box 1999
Sudbury
Suffolk CO10 6FS

Tel: 01787 881165 fax: 01787 313995

HSE Infoline: tel 0541 545500

HSE Bookfinder: Subscription service from HSE Books

HSE local offices around the UK: find their number in your local Telephone Book

Department for Employment, Transport and the Regions (DETR)

Great Minster House
76 Marsham Street
London SW1P 4BR

Website: *www.detr.gov.uk/hsw/index.htm*

Department for Education and Employment (DfEE)

For information on employing people with disabilities:

DDA Helpline: 0345 622 633 or 0345 622 644

Department for Trade and Industry (DTI)

DTI Consumer Safety Unit
CA3a, 4th Floor
1 Victoria Street
London SW1H 0ET

Tel: 0207 1215 0359 fax: 0207 1215 0357

For information on publications, contact:

DTI Publications Orderline: 0870 1502 500

Department of Health

Department of Health
Wellington House
133–155 Waterloo Road
London SE1 8UG

Tel: 0207 1972 2000

Department of Social Security (DSS) regarding sick pay or maternity pay

DSS Advice Service: 0345 143 143

Local Authority Departments

Many health, safety or fire issues are dealt with at a local level through the Local Authority. The relevant number should be in the local Telephone Book for:

- Environmental Health Office
- Trading Standards Office
- Planning Department
- Fire Service

Most Local Authorities have some information or guidance for small firms in their area, for example

Test Valley Borough Council produce *A Handy Guide to Health & Safety*
Dundee City Council produce *Risk Assessment for the Smaller Food Business*.

17.2 Industry groups

There is likely to be an Industry Representative Group for your type of business, or a general representative group such as:

(i) Federation of Small Businesses, 2 Catherine Place, Westminster, London SW1E 6HF
(ii) Forum of Private Business, Ruskin Chambers, Drury Lane, Knutsford, Cheshire WA16 6HA
(iii) Confederation of British Industry, Centre Point, 103 New Oxford Street, London WC1A 1DA
(iv) A local Chamber of Commerce – see local Telephone Book.

17.3 Health and safety professionals

(i) Institution of Occupational Safety & Health (IOSH), The Grange, Highfield Drive, Wigston, Leics LE18 1NN
(ii) BMA, BMA House, Tavistock Square, London WC1H 9JP
(iii) Faculty of Occupational Medicine, Royal College of Physicians, 6 St Andrews Place, Regents Park, London NW1 4LB
(iv) Employment Medical Advisory Service (EMAS): contact HSE 0541 545500 for local office
(v) NHS Pensions Agency, Injury Benefits Manager, 200–220 Broadway, Fleetwood, Lancs FY7 8LG
(vi) The National Aids helpline: 0800 567 123
(vii) You can also contact your local GP for information about local occupational therapists or occupational hygienists.
(viii) British Safety Industry Federation, tel: 01745 585600

17.4 Other bodies and organizations with an interest in fire, health and safety

(i) British Safety Council, National Safety Centre, 70 Chancellors Road, London W6 9RS
(ii) Royal Society for the Prevention of Accidents (ROSPA), Edgbaston Park, 353 Bristol Road, Birmingham B5 7ST
(iii) Loss Prevention Council/Fire Protection Association, Melrose Avenue, Boreham Wood, Herts, WD6 2BJ
(iv) British Fire Protection Systems Association, 4th Floor, Neville House, 55 Eden Street, Kingston upon Thames, Surrey, KT1 1BW
(v) Arson Prevention Bureau, 51 Gresham Street, London, EC2V 7HQ
(vi) Association of British Insurers, 51 Gresham Street, London, EC2V 7HQ
(vii) British Standards Institute (BSI), 389 Chiswick High Road, Chiswick, London W4 4AL
(viii) Trades Union Congress (TUC), Congress House, Great Russell Street, London WC1B 3LS
(ix) ACAS: web site *www.acas.org.uk*, or phone nearest local office
(x) Health & Safety Agency in Ireland (HAS), 10 Hogan Place, Dublin 2

17.5 Media and other providers of guidance or support

(i) Butterworths Tolley *Health & Safety at Work Journal*, Anne Boleyn House, 9–13 Ewell Road, Cheam, Surrey SM3 8JT
(ii) *Management of Occupational Health, Safety and Environment Journal*, 34 Warwick Road, Kenilworth, Warks CV8 1HE
(iii) IOSH *The Safety & Health Practitioner Journal*, Paramount Publishing Ltd, Paramount House, 17–21 Shenley Road, Borehamwood, Herts, WD6 1RT
(iv) *Financial Times* guide *Business Health & Safety*, annually
(v) Croner Publications, Croner House, London Road, Kingston upon Thames, Surrey KT2 6BR
(vi) Gee Publishing Ltd, 100 Avenue Road, Swiss Cottage, London NW3 3PG
(vii) Tolley Publishing, Anne Boleyn House, 9–13 Ewell Road, Cheam, Surrey SM3 8JT
(viii) *Directors' Briefings* from Business Hotline Publication Ltd, 268 Lavender Hill, London, SW11 1LJ

17.6 Training providers

There are likely to be many locally based providers of training in health and safety or fire risk management, whether through Colleges and Universities, Business Links or as Consultants. Others include:

(i) British Red Cross (for first aid training), Commercial Training Centre, 163 Eversholt Street, London NW1 1BU
(ii) ROSPA, Edgbaston Park, 353 Bristol Road, Birmingham B5 7ST
(iii) Industrial Relations Services Training, Lincoln House, 296–302 High Holborn, London WC1V 7JH
(iv) Chartered Institute of Environmental Health (CIEH), Chadwick Court, 15 Hatfields, London SE1 8DJ
(v) Rapid Results College, Tuition House, 27–37 St George's Road, London SW19 4DS

17.7 Publications

HSE provide a range of free leaflets, such as

- INDG 259 – Health & Safety in Small Firms
- MISC 130 – Good Health is Good Business
- INDG 232L – Consulting employees on Health & Safety
- HSE 31 – RIDDOR explained
- INDG 213 – 5 Steps to information instruction and training
- INDG 136(REV1) – COSHH A brief guide to the regulations
- INDG 272 – Health Risk Management: a Guide to working with Solvents
- INDG 273 – Working safely with solvents

- INDG 160L – Maintaining portable electrical equipment in offices
- INDG 164L – Maintaining portable electrical equipment in hotels
- INDG 173L – Officewise
- INDG 36L – Working with VDUs
- INDG 226L – Homeworking
- INDG 199L – Managing vehicle safety at the workplace
- INDG 215L – First Aid
- INDG 297 – Safety in gas welding
- CDM Regulations – How the Regulations affect you

Contact HSE Books 01787 881165 to order or for a copy of the catalogues for free and priced publications. They also produce a range of other guidance and materials, including training videos and packs.

17.8 Other contacts

Most providers of information have a website on the Internet, including government departments and manufacturers who often have a great deal of free guidance available. Suppliers of fire fighting or security equipment should also be able to provide specific guidance for your own needs.

Index